SEP 28 1962

MAGNETIC CONTROL
OF
INDUSTRIAL MOTORS

PART 2   *Alternating-Current
         Motor Controllers*

MAGNETIC CONTROL
OF
INDUSTRIAL MOTORS

Gerhart W. Heumann
*Industrial Engineering Section,
General Electric Company*

Part 2
*Alternating-Current
Motor Controllers*

John Wiley & Sons, Inc.
New York and London

COPYRIGHT © 1961 BY JOHN WILEY & SONS, INC.

All rights reserved. This book or any part thereof must not be reproduced in any form without the written permission of the publisher.

LIBRARY OF CONGRESS CATALOG CARD NUMBER: 61-11593
PRINTED IN THE UNITED STATES OF AMERICA

to MARIANNE BOBBY HEUMANN

PREFACE

Expansion of industrial activities, mechanization of manufacture, and automation of industrial processes have greatly increased the use of motors and their control in industry. The application of control and motors has become recognized as an important branch of electrical engineering. It has become the subject of an increasing number of courses in engineering schools.

Control implies the control of the flow of energy in a predetermined manner. It may be the control of the flow of electric energy in a power system. It may be the control of heat in a chemical process or of combustion in an engine. It may be the control of the trajectory of a guided missile.

Industrial control, by generally accepted definition, pertains to apparatus applied to the utilization of electric power. It denotes the obtaining of specific performance from industrial equipment. In the majority of cases, it concerns the control of a motorized drive. The motor provides the mechanical force to move the drive, and the control equipment determines what the motor is to do. Equipment which governs the distribution of power and the protection of the power-supply system belongs in the classification of switchgear.

When electric motors were first introduced, simple manual switches were used to start and stop the motors, and fuses were the only protective devices available in those days for protecting the motor and the power system. Progress was initially along the lines of improving the manual control devices, increasing reliability, flexibility, and make and break performance. But each starting and speed-control operation had to be performed by hand, as an operator moved a manual switching device from one position to another. Control of large motors required great physical effort. To supervise a drive, the operator had to be continually alert to watch indicators as he adjusted motor performance to the drive requirements. Interlocking or sequencing of related drives, starting, stopping, and speed adjustment as a function of position or other process quantities could not be accomplished.

An important advance in the control art was the introduction of electromagnetic contactors and relays, that is, devices actuated by electromagnets and requiring for their actuation only a small fraction of the power which can be switched on the main contacts. Thus electromagnetic contactors and relays are power amplifiers. Switching no longer requires physical effort by the operator; and the low power level needed to actuate contactors and relays permits the introduction of control-circuit devices which respond to manual commands by an operator, to impersonal (or automatic) commands as a function of machine position, or to some measurable drive quantity. The principal advantages of magnetic control, as compared with manual control, are greater flexibility, greater safety, and convenience to the operator.

If manual control is used for a large motor, the physical location of the controller presents a problem. Locating the controller close to the motor permits the shortest runs of heavy power leads. However, from the operator's point of view, this may be a very poor location, making it difficult for him to observe and operate the machine. Placing the controller a distance away from the motor requires long runs of heavy power cables. Magnetic control permits installing the power contactors close to the motor, whereas the actuating control devices, requiring only small-size control wiring, can be located away from the motor in a position most convenient to the operator.

Many industrial applications, such as automated manufacturing lines, continuous processing lines, movable bridges, and conveyor trains, involve the use of a number of motors that must be controlled by a single operator at a work station, from which he can observe and supervise all parts of the installation. With magnetic control, a large number of control-circuit devices can be consolidated in a small space, greatly reducing the physical and mental effort necessary to operate a large and complex piece of machinery. With magnetic control, it is easy to interlock various operations of a multimotor drive so that different motions can be performed only in their proper sequence, thereby preventing false starts which are a hazard to the life and limbs of the operator as well as to the machinery.

Many industrial processes require automatic operations of motors as a function of quantities measured by delicate instruments. The small amount of control power required to actuate contactors can be handled by sensors or pilot devices responsive to operating signals. It would be impractical to design such devices with contacts capable of handling power circuits. As an example, consider an electric furnace. A delicate temperature-sensitive instrument, with the help of contactors, maintains furnace temperature much more accurately than manual control ever could.

Magnetic control removes the operator physically from power circuits and contacts which open power arcs. The physical remoteness minimizes the danger to operators of coming into accidental contact with live parts or being exposed to power arcs and flashes. When high-voltage motors are started and stopped frequently, magnetic contactors are the only means of controlling such motors without hazard to the operating personnel.

An important advantage of magnetic control is the ease with which the performance of the motor and the driven machinery can be adjusted and maintained, independent of the skill of the operator. Current and torque peaks can be limited, thus avoiding unnecessary wear and tear which may be caused by a manual controller in the hands of an unskilled operator. The result is a reduction in maintenance expense required for the motor and the controller.

To many people today, the term "control" means automatic control, that is, sophisticated systems involving sensing of process quantities, automatic feedback controllers, data processing, programming, and so forth. Although these highly refined control systems involve advance technologies which have progressed rapidly since World War II, basically they utilize magnetic control devices as the power actuators which open, close, and modulate motor power circuits; that is, magnetic control is one of the building blocks of an automatic control system. Regulators, feedback loops, sensors, computers, programmers, and other controls ultimately actuate magnetic devices which perform the power switching functions.

Magnetic devices have been developed to such a degree that they are the most versatile and economical type of power switching devices available to the control designer. Their performance and life expectancy are reasonably high in relation to their cost. Improved materials and better design will bring about progress in the existing lines of devices, without the need to introduce radically new concepts. Some day magnetic switching devices may be superseded by static devices, which are now feasible and economical in control circuits handling power at very low levels. It can be expected that eventually static switching devices capable of switching motor power circuits will become available at reasonable cost. However, it will be quite a few years yet before magnetic contactors and relays are a thing of the past; meanwhile, magnetic control will continue to be an important factor in the total field of motor control.

Conventional magnetic control is treated in these books. It is a sufficiently broad subject to deserve treatment by itself. Automatic control, analysis and synthesis of feedback control systems, programming, computing, and related subjects warrant separate treatment and are therefore not treated, except incidentally to point up the interrelationship between conventional magnetic and automatic control.

Control engineering comprises two major phases. One is the design of devices and circuit components of which controllers are built. The other is the design of controllers and equipments. This latter phase of control engineering requires the study of application engineering problems, the design of circuits to solve the application requirements, the selection of devices necessary to realize the chosen circuitry, and finally the physical layout of a controller which contains the devices, properly connected, in a mechanical arrangement suitable for the specific conditions of the application. This phase of control engineering is treated in these books.

Changes in the art of control occur in rapid succession, either as a result of technological developments in industry or the application of new knowledge and materials in electrical engineering. *Magnetic Control of Industrial Motors* makes use of the basic material contained in earlier books and presents new material reflecting recent changes in the design of control devices and equipments. Because of the many innovations in industrial control, it was decided to publish this material in three parts, since a single book would become too bulky. The three parts are closely related, yet each is largely a self-contained unit which presents a certain area of the motor control field. Together, they present an integrated picture of the whole field of a-c and d-c industrial motor control.

Three-phase a-c power is available today in all industrial areas; indeed, the days of the general d-c distribution system are just about over. Alternating-current motors, which are simple, reliable, and comparatively inexpensive, are the work horses of industry. They are used wherever motorized drives are single units and where drive requirements can be satisfied by the comparatively simple characteristics of a-c motors. Since every industrial establishment contains a-c motors, a knowledge of their performance and control is one of the basic requirements of an electrical engineer working on industrial problems. Parts 1 and 2 are devoted to a-c motor control problems. Part 1 deals with the characteristics of a-c control devices and their assemblage into complete controllers. Part 2 covers specific controllers for a-c motors and the general problem of protection of motors and their branch circuits.

Direct-current power distribution for general use is uneconomical, compared with a-c power distribution. Hence the use of d-c motors for general-purpose drive applications is decreasing. Their principal use is for special-purpose drives requiring the superior abilities of d-c motors in regard to the control of speed, torque, power, rate of acceleration and retardation, dynamic braking, and so forth, or requiring the integration of several motors into a coordinated drive. Local d-c distribution systems are still found in areas needing a large amount of d-c power for motor drives, as in steel mills. In areas served by a-c power distribution systems, d-c power is obtained by conversion of the general-purpose a-c power available in the plant at the spot where d-c power is needed for a specific complex drive. Control for d-c motors is treated in Part 3.

Industrial control engineering is a science and an art. Given the mechanical characteristics of a drive, the performance characteristics of a motor and the layout of a control equipment, the control engineer can calculate rigorously the performance which can be expected of the equipment. He can predict how fast the driven machine will run, within what time it will accelerate, how it will decelerate, how the speed will react to torque changes, and other performance factors.

To be sure, it is not always possible to make such a prediction. It may be impossible, for example, to forecast accurately the performance of a drive because of difficulties in obtaining sufficiently accurate mechanical data. Friction may change considerably, depending on loading, drive adjustment, seasonal changes in ambient temperature, and other factors. It may also be impossible to obtain accurate motor data, and a general knowledge of motor characteristics may be all that would be available. In such cases it becomes the task of the control engineer to use proper judgment and, based on his experience and knowledge of the application, to design controllers that possess sufficient flexibility for adequate performance, even if the mechanical characteristics of the drive vary over a wide range.

Economic factors have an important effect on controller design. Control equipment is built for one purpose: to be sold to a prospective user. Experience and judgment influence the decision as to what type of control to apply. The control engineer must know which features are absolutely necessary and which features, though desirable, could be dispensed with. It then becomes a question of whether the cost of the desirable features is economically justified. It is good judgment to offer equipment which, although it may not contain all possible refinements, is attractive in cost.

Compared with switchgear, control becomes obsolescent more rapidly. Industrial processes change; industrial equipments may be replaced before they become worn out in order to effect production economies. Hence the tendency is to keep control costs down to a level commensurate with adequate performance, with less emphasis on appearance and long service life than one would customarily expect, for instance, of switchgear apparatus.

The language of the control engineer is an intricate system of electrical symbols and diagrams. Part 1 opens with a presentation of symbols and discussion of these symbols and diagrams. Chapters 2, 3, 4, and 5 introduce a-c devices and components used as parts of controllers—the building blocks of which controllers are constructed.

Chapter 6 deals with solenoids, Thrustor * mechanisms, and brakes, which are electrically actuated components performing mechanical work. The reason for treating these components here is that they have to be integrated into the controller design.

Once the circuits have been designed and the required components for realizing the circuitry and obtaining the desired motor performance selected, the concluding step in designing the controller is the mechanical assembly of devices into a complete unit. Chapter 7 is devoted to the mechanical design of controllers. To protect personnel against accidental contact with live parts and to protect the equipment against ambient conditions, many controllers are enclosed. Function and design of enclosures are included in this chapter.

Controllers, as is any type of machinery, are subject to wear. To obtain satisfactory service life, control equipment must be maintained. The concluding Chapter 8 gives general recommendations for the servicing and maintenance of control equipment.

Part 2 starts out with a discussion of characteristics of mechanical drives powered by electric motors. A knowledge of these drive characteristics is essential as a guide in selecting the type of motor best suited for a particular application and the type of controller needed to obtain the required performance of the selected motor.

Chapters 2, 3, and 4 present circuitry and selection of controller components for commonly used industrial a-c motors, such as squirrel-cage, wound-rotor, and synchronous motors. Since the control engineer must possess a basic knowledge of motor characteristics, particularly speed-torque curves, each chapter starts out with a discussion of the

* Registered trade mark of the General Electric Company.

fundamental motor characteristics. Control and power circuits to obtain the desired motor performance are discussed and illustrated by basic circuit diagrams. Pointers are given on selection of control devices.

A recent development in the control art are static switching components which permit obtaining considerably longer life and greater reliability than can be obtained with magnetic contactors and relays. Chapter 5 discusses static switching and the concept of logic functions, an important step in simplifying circuit design for complex control systems.

In addition to starting and stopping motors, and modifying their speed-torque characteristics, all controllers include protection of the motor against overloads. Furthermore, many controllers include short-circuit protection. The controller is part of the motor branch circuit which is connected to the power distribution system. To safeguard the power distribution system against damage due to faults, short-circuit protection is necessary. This protection may be supplied as part of the controller, or it may be supplied separately. To obtain optimum safe operation of the complete power system, from the distribution switchboard to the point of utilization, proper coordination of overload and fault protection has to be considered in the controller design. Therefore, Chapters 6 and 7 are devoted to the problems of a-c motor and branch-circuit protection.

Part 3 deals with the control of d-c motors. Although many control devices can be used on a-c as well as d-c controllers, there are certain characteristics which are unique to d-c devices. Hence the first two chapters introduce d-c contactors and relays as well as d-c solenoids and brakes. Chapter 3 treats the performance characteristics of d-c series motors which are used primarily on constant-voltage power systems for driving heavy intermittent material-handling equipments, such as cranes and hoists. Chapter 4 deals with shunt motors operated on a constant-voltage power supply. Although the general-purpose d-c distribution systems are shrinking, such power systems are still in existence in heavy mill areas, and d-c shunt motors are often given preference for mill auxiliary drives because of their greater flexibility compared to a-c motors.

The highest degree of d-c shunt-motor performance is obtained by the adjustable-voltage system forming the subject of Chapter 5. This method of control is suitable for the most complex drive system. Direct-current power is not obtained from a general-purpose d-c power supply, but it is produced from an individual conversion unit, the output of which is controlled to suit the individual drive system. The classical conversion unit, and still the preferred one for large drives, is the d-c shunt generator, although static rectifying untis are gaining acceptance in the range of small- and medium-size drives. Regulating exciters, which serve to control and regulate generator and motor fields, are dealt with in Chapter 6. They form the connecting link between the power units and automatic control, in which feedback loops may be introduced. Although this book does not treat automatic control systems in depth, examples are given of the way in which automatic feedback control can be introduced in an adjustable-voltage system. Concluding Chapter 7 describes packaged drives in which power conversion equipment and control are coordinated and assembled in a single unit, the "package" being engineered and sold as a single product.

Whenever applicable, the "per unit" system is used to represent the performance characteristics of motors and controllers. In this system various quantities are expressed as a ratio to a base quantity which is taken as unity. Base quantities have to be defined in each case. Usually, they are the rated or normal quantities. As an example, rated current of a motor is assumed to be its base current and rated torque its base torque. To express the speed of a-c motors, it is convenient to assume synchronous speed as base speed. To interpret any presentation in per unit properly, care must be taken to determine the base quantities that apply.

Representing performance data in per unit has the advantage that such data are generally applicable, regardless of the size or rating of the machine for which they have been computed. To translate per unit data into numerical data applying to a specific machine, it is necessary only to multiply the per unit quantities by the numerical value of the base quantity. In technical literature, performance data are expressed quite often in percentages. The per unit system and the percentage system are equivalent. Unity in the per unit system corresponds to 100 per cent in the percentage system. The per unit system is more convenient for carrying out calculations, since per unit quantities can be introduced directly into formulas, whereas calculations with percentage quantities require divisions by 100 or powers of 100.

Formulas which cannot be used for per unit calculations and which should be used only for numerical calculations are followed by the symbol ●. Formulas which are correct only on a per unit basis and cannot be used for numerical calculations are marked by the symbol ■. All other formulas can be used for both per unit and numerical calculations. See pages 1, 9 and 83 for examples of each kind of formula.

Each chapter closes with a bibliography listing books, papers, and articles which are obtainable from reference libraries. Entries were selected for their practical applications to the subject on hand, rather than for mathematical and highly technical analysis.

In order to make these books useful as a teaching text, appropriate problems are listed at the end of those chapters in which the subject matter is suited for problems to be given to students. These problems are practical problems which may be encountered by a control design engineer in his daily work.

Since the publication of earlier books, there have been many changes in industrial practices affecting the design of motors and control apparatus. It is the aim of these books to reflect these changes through the inclusion of considerable new material and the elimination of obsolete designs.

A number of comments have been received, particularly from teachers who expressed a preference for having the performance of a certain type of motor and its control treated in the same chapter. As a result, the treatment of motor characteristics and their control has been rearranged.

Changes in industrial practices have resulted in revisions in existing industrial standards, especially NEMA Standards for Industrial Control and American Standard C19.1 for Industrial Control. The latest issues of recognized industrial standards have been considered in the preparation of statements, tables, and diagrams.

I wish to emphasize the role of teamwork without which the industrial control art could not progress as rapidly as it has in the past. I wish to express sincere appreciation to my associates in the General Electric Company, and to other industrial control manufacturers who are members of the NEMA Industrial Control Section, for having placed material at my disposal and assisting me in the preparation of these books.

GERHART W. HEUMANN

Schenectady, N. Y.
March 1961

CONTENTS

1	DRIVE CHARACTERISTICS	1
2	CONTROL OF SQUIRREL-CAGE MOTORS	8
3	CONTROL OF WOUND-ROTOR MOTORS	81
4	CONTROL OF SYNCHRONOUS MOTORS	176
5	LOGIC FUNCTIONS AND STATIC SWITCHING	215
6	MOTOR PROTECTION	244
7	MOTOR BRANCH-CIRCUIT PROTECTION	292
	INDEX	329

1 DRIVE CHARACTERISTICS

To design a motor controller, it is necessary to know the characteristics of the motor as well as the those of the driven load. Usually, an application engineer gathers the necessary data on the driven machine, selects the size and type of motor to be used, and establishes the desired performance characteristics. It then becomes the task of the control engineer to select equipment and design circuits such that motor performance matches the requirements of the drive and proper protection is provided for abnormal conditions.

SPEED-TORQUE CURVES

Motor performance can be most easily visualized by plotting speed versus torque. In order to apply a motor to a machine, it is necessary to know the torque which the motor is required to deliver at its shaft and the speed at which the machine must run to obtain the desired output of work. Motors are rated in horsepower and speed. From these data, the rated full-load torque can be determined as follows:

$$\text{Torque in lb-ft} = \frac{\text{rated hp} \times 5252}{\text{full-load rpm}} \qquad (1) \bullet \ {}^*$$

To compare the performance of various types and sizes of motors and to apply results calculated for one motor to other motors of similar characteristics, it is convenient to express speed and torque in per unit. If rated full-load torque and speed are considered base quantities or unity, motor performance at partial load and partial speeds can be expressed in per unit of rated full-load values.

When motor speed is plotted versus motor torque, the performance of the drive under varying loads can be deducted. It is also possible to predict how changes in the motor circuit, as obtained by the con-

---

* Formulas which are correct only for numerical calculations and cannot be used for per unit calculations are followed by the symbol ●. Formulas which are correct only on a per unit basis and cannot be used for numerical calculations are followed by the symbol ■ (see page 9). All other formulas in this book hold true for both per unit and numerical calculations.

troller, will affect machine performance. Conversely, if it is known what performance is desired of the machine, a study of the speed-torque curves will aid in the selection of a suitable control circuit. In like manner, the characteristics of loads can be represented by plotting speed versus torque, to indicate changes in the required motor duty corresponding to changes of load on the drive.

In most cases, torque represents a load imposed on the motor by extraneous influences, and the question arises: what speed corresponds to a given load or torque. Therefore, in plotting speed-torque curves, it is customary to plot torque as the abscissa and speed as the ordinate. These two coordinates determine four quadrants in which speed and torque may be positive or negative. The quadrants represent all possible operating conditions of a motor or a load, as illustrated in Fig. 1-1.

In the upper right-hand quadrant 1, both motor torque and speed are positive. Using a motor driving a hoist as an example, this quadrant represents the condition that the motor hoists a load. Motor torque overcomes gravity; the motor delivers power to the load. An equivalent condition exists when the motor drives a friction or inertia load, or a combination of both, in a given direction, which is arbitrarily called positive.

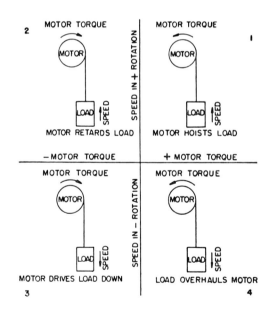

Fig. 1-1 The four quadrants of motor performance.

The upper left-hand quadrant 2 represents the operating condition that speed is still positive but the torque is negative. The motor develops a torque which is opposed to the load torque and retards the load. Expressed differently, the motor develops a braking torque. When a motor drives an inertia load, quadrant 2 represents the condition that the motor develops a torque opposing the overhauling load torque due to the kinetic energy of the inertia load. The motor absorbs energy from the load, which is either dissipated as ohmic loss or fed back to the power supply.

In the lower left-hand quadrant 3, both torque and speed are negative. In a hoist drive, this condition exists when the motor accelerates its own inertia and the load is accelerated by gravity in the lowering direction. Motor torque is in the same direction as the load torque, gravity pulling the load downward. A corresponding condition exists when the motor drives a friction or inertia load in a direction of rotation opposite to the one existing in quadrant 1. Power is delivered by the motor to the load. Quadrants 1 and 3 are equivalent, except that direction of rotation is reversed.

The lower right-hand quadrant 4 indicates that motor torque is positive, while speed is negative. In a hoist, for example, motor torque opposes the load torque, which is the pull of gravity. While the motor attempts to hoist the load, the load overhauls the motor, so that actual movement is in the lowering direction. The motor torque restrains the load; that is, the motor develops a braking torque. In describing hoist performance, this quadrant is often called the braking-lowering quadrant. Energy is absorbed by the motor from the load and dissipated as ohmic loss, or fed back to the power system. When a motor drives an inertia load, motor torque opposes the overhauling load torque due to the kinetic energy of the inertia. Quadrants 2 and 4 are equivalent, except for reversal of the direction of rotation.

From the motor designer a set of characteristic curves can be obtained which describe the performance of the motor in quadrant 1, that is, with torque and speed positive. From these motor curves, performance in the other quadrants can be calculated.

LOAD CHARACTERISTICS

The two basic types of loads, as illustrated in Fig. 2-1, are friction loads and fan loads. In Fig. 2-1A, the relation between speed and torque is given. With a friction load, torque is constant over the speed range. Typical of this type of load is a crane hoist. A load, suspended on a crane hook, is subjected to the constant pull of gravity. Assuming

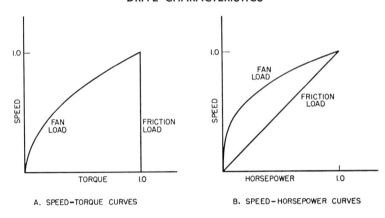

Fig. 2-1 Basic load characteristics.

a cylindrical hoist drum, the torque exerted on a motor shaft would also be constant. Another example of a friction load is that encountered on cutting-type machine tools. Controllers designed for constant-torque applications are sometimes designated as controllers for machine-tool duty. In Fig. 2-1B, the relation between horsepower and speed is plotted. With a constant-torque load, horsepower varies proportionately with speed.

Another basic type of load is that encountered in transporting a viscous fluid. Since a large portion of energy is utilized for accelerating the fluid, torque varies as the square of the speed. Typical examples are fans, centrifugal compressors, and centrifugal pumps. This type of load is generally designated as "fan load," and its speed-torque characteristic is illustrated in Fig. 2-1A. Horsepower varies as the cube of the speed, as indicated in Fig. 2-1B. While the curves given in Fig. 2-1 describe the theoretical characteristics of a pure fan load, actual fans, pumps, and compressors deviate somewhat from these theoretical curves. Friction of mechanical parts has to be considered. Friction in ducts and pipelines also causes some deviation from the idealized fan-load curves. If close control or regulation of speed is necessary, the actual characteristic should be obtained from the machinery manufacturer. This is particularly important when the actual load falls between the fan load and a friction load. An example is a pneumatic conveyor. The compressor unit as such is essentially a fan load, to which is added the friction load due to the transport of the suspended material in the conveyor ducts.

ACCELERATION AND RETARDATION

To accelerate a load, the motor must develop a torque in excess of the load torque. This excess torque accelerates the inertia of the moving masses, as expressed by the Wk^2 of the motor armature and the connected masses of the driven load. If the load and the motor torque characteristics are known, the time required to accelerate from one speed to another can be determined from the following formula:

$$\Delta t = \frac{Wk^2 \times \Delta S}{308 \times \Delta T} \qquad (2)$$

where t = accelerating time in seconds
Wk^2 = moment of inertia in lb-ft^2
ΔS = difference in speed in rpm
ΔT = average accelerating torque in lb-ft

A major portion of the total moment of inertia is generally the Wk^2 of the motor armature, which can be obtained from motor handbook listings or, for special motors, from the motor designer. The moment of inertia of the driven machinery must be obtained from the machinery builder. Since calculations of acceleration performance are usually made in terms of motor speed, the moment of inertia of machinery parts which rotate at a speed different from the motor speed must be reduced to an equivalent moment of inertia at motor speed. It is important that this conversion be made on machinery which is geared to the motor. If part of the drive, rotating at a speed of S_1 rpm, has a moment of inertia $W_1 k_1^2$, the equivalent moment of inertia at speed S is

$$Wk^2 = W_1 k_1^2 \left(\frac{S_1}{S}\right)^2 \qquad (3)$$

The load may include masses having linear motion. For instance, part of the load on the motor shaft may be the mechanical load suspended from the hook of a crane hoist. It may be the weight of a crane bridge or trolley, or it may be the reciprocating bed of a planer. The equivalent moment of inertia of the linearly moving masses is

$$Wk^2 = W \left(\frac{V}{2\pi S}\right)^2 \qquad (4)$$

where W = weight of moving mass in pounds
V = linear velocity of moving mass in feet per minute
S = motor speed in rpm corresponding to velocity V

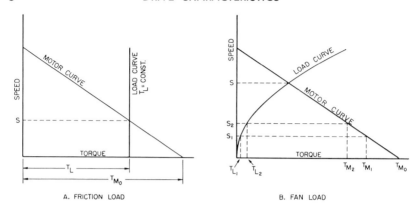

Fig. 3-1 Determination of average accelerating torque.

Equation 2 can also be used to determine the time required for retarding a load. In place of the average accelerating torque, the average braking torque is inserted in equation 2 as ΔT. If the load is retarded by a mechanical brake, the constant brake torque is used as the retarding torque ΔT. If electric braking is used, the motor characteristic giving braking torque as a function of speed must be known in order to determine the amount of motor torque available for retarding the load.

Average accelerating torque ΔT is the amount of motor torque available as a surplus over the torque which is necessary to overcome the load torque. Figure 3-1 illustrates how the average accelerating torque can be evaluated. Figure 3-1A indicates the conditions which exist for a friction load whose speed-torque curve can be described as straight line T_L = const. As far as motor behavior is concerned, it is assumed that the motor speed-torque curve is a straight line between motor starting torque T_{M0} at S = 0 and a given no-load speed at $T = 0$. If the motor is started from rest, it will accelerate to a speed S, at which motor torque equals the load torque. From Fig. 3-1A, it can readily be seen that the average accelerating torque available for accelerating the load from rest to speed S is $\Delta T = \frac{1}{2}(T_{M0} - T_L)$.

If the load is not constant over the speed range, but its characteristic is given by a curve, as illustrated in Fig. 3-1B, the operating speed to which the motor accelerates is determined by the intersection between the motor curve and the load curve. Average accelerating torque and time could be calculated by purely mathematical analysis if the load curve were a true parabola. However, actual load curves generally deviate from a pure mathematical function, even in the case of an ordinary fan, because of mechanical friction and turbulence due to mechanical parts which may interfere with the fluid flow. Acceleration performance for any shape of load curve may be determined point by point in the following manner: The total speed range is subdivided into short speed intervals within which the load curve may be replaced by short straight lines. For acceleration from standstill to speed S_1, $\Delta S = S_1$. The average load torque is $\frac{1}{2} T_{L1}$, and the average motor torque is $\frac{1}{2}(T_{M0} + T_{M1})$. Hence the average accelerating torque is $\Delta T = \frac{1}{2}(T_{M0} + T_{M1} - T_{L1})$, and accelerating time Δt_1 may be calculated from formula 2. To calculate acceleration to speed S_2, the speed increase is $\Delta S = S_2 - S_1$. Between S_1 and S_2 the average load torque is $\frac{1}{2}(T_{L1} + T_{L2})$, and the average motor torque is $\frac{1}{2}(T_{M1} + T_{M2})$. The average accelerating torque is then $\frac{1}{2}(T_{M1} + T_{M2} - T_{L1} - T_{L2})$, and the accelerating time interval Δt_2 may then be calculated from formula 2. This procedure is repeated until speed S is reached. If the motor curve is not a straight line, the same point-by-point procedure may be followed, replacing the motor curve by short straight lines between speed points. The same point-by-point procedure is also used to determine the average retarding torque due to electric braking, when motor braking torque versus speed is known.

Accelerating time calculated according to formula 2, with an average accelerating torque, yields results which are of satisfactory accuracy for practical purposes as long as the accelerating torque is not dropping to a value less than about one quarter of its initial value. If the accelerating torque drops to zero or nearly zero, accelerating time calculated with an average accelerating torque may be up to one quarter shorter than the actual accelerating time. With diminishing accelerating torque, rate of change of speed deviates considerably from the constant value implied in formula 2, when using the average accelerating torque. This deviation accounts for the difference between actual and calculated accelerating time.

BIBLIOGRAPHY

1. T. C. Johnson (Editor), *Electric Motors in Industry,* Chapter 9, John Wiley and Sons, New York, 1943.
2. R. P. Ballou, "Load Analysis as a Basis for the Selection of Motor and Drive," *Electrical Manufacturing,* May 1948.
3. W. R. Harris, "Mechanics of Applying Electric Motors," *Machine Design,* November 28, 1957.
4. R. W. Jones, *Electric Control Systems,* Chapter 2, John Wiley and Sons, New York, 1953.

2 CONTROL OF SQUIRREL-CAGE MOTORS

Hardly ever will a control engineer be called upon to design an electric motor. To apply control to a motor, he does not have to lay out its windings, calculate its flux, or determine its physical dimensions. However, to coordinate a motor and its control to best advantage, the control engineer has to know how a given motor works and how its characteristics can be influenced by the control to obtain a desired performance. Chapters 2, 3, and 4 will therefore include introductory sections devoted to a brief discussion of the characteristics of various types of a-c motors, how they work, and what they can do.

The most widely used types of a-c motors in industry are polyphase induction motors of the squirrel-cage and wound-rotor type, and synchronous motors. Since single-phase motors and a-c commutator-type motors, although of great importance for appliances and transportation work, are not frequently used in industrial applications at the present time, they are not treated in this book.

PRINCIPLES OF INDUCTION-MOTOR OPERATION

An induction motor consists of a stationary frame or stator which contains the primary winding, the design of which is determined by the number of phases and the number of poles. The rotating member, or rotor, contains the secondary winding, which may consist of bars short-circuited by rings at the end. A motor of that type is called a "squirrel-cage motor" because of the resemblance of the rotor winding to a squirrel cage. A squirrel-cage winding is short-circuited in itself, and no external connections may be made to it. Or, the rotor winding may be a distributed polyphase winding, the phases of which are terminated at slip rings. Such a motor is called a "wound-rotor motor." By means of external connections to the slip rings, the electric characteristics of the secondary rotor circuit may be varied, thus affecting the performance of the motor.

An induction motor, with its stator connected to the polyphase power

PRINCIPLES OF INDUCTION-MOTOR OPERATION

supply and the rotor at standstill, acts as a transformer. Current flows through the primary or stator winding, the magnitude and phase angle of which are determined by the impedance of the motor. That component of current which lags line voltage by 90 degrees determines the flux set up in the stator. It is called the magnetizing current and is substantially independent of the load on the motor.

In a polyphase motor, the flux produced by each phase current reaches its peak in the same sequence as the phases of the applied line voltage. The total flux produced by the various phases acting in combination rotates in space in a direction determined by the phase sequence of the applied line voltage. The speed at which the flux rotates in space is equal to 360 electrical degrees per cycle, or an angle of 360 degrees divided by the number of pairs of poles of the stator winding. It can be expressed in revolutions per minute:

$$\text{Speed of flux rotation in rpm} = \frac{60 \times \text{frequency}}{\text{pairs of poles}} \quad (5) \bullet$$

This rotating flux intersects the conductors on the rotor and, at standstill, a secondary voltage is generated, the magnitude of which depends on the turn ratio between the stator and the rotor windings. The generated rotor voltage causes current to circulate in the rotor winding, its magnitude and phase angle depending on the impedance of the rotor circuit. That component of current in phase with the rotor voltage interacts with the flux and develops a torque which causes the rotor to turn in the same direction as the flux rotates.

Speed in rpm at which the flux rotates in space is the synchronous speed, which is base speed or unity. The difference between synchronous speed and actual speed of the motor is called "slip," usually expressed in per unit of synchronous speed:

$$S = 1 - s \quad (6) \blacksquare$$

where S = motor speed in per unit
s = slip in per unit

Since synchronous speed is proportional to line frequency, slip can be expressed in cycles per second, thus:

$$s(\text{cycles}) = \text{line frequency} \times s \quad (7) \bullet$$

This is the frequency with which the rotor conductors intersect the rotating flux. It is the frequency of the voltage induced in the rotor conductors. At standstill, slip is unity. When the rotor turns in the direction of flux rotation, which corresponds to normal motor operation, slip is less than unity. If the rotor is driven by an external force against flux rotation, slip is greater than unity. If the rotor is driven by an external force in the direction of flux rotation, but at a speed above synchronous speed, slip is negative.

When an induction motor is started from standstill and its speed increases, slip decreases. Likewise, frequency and magnitude of the rotor voltage decrease. If a motor could be built without any losses, it would accelerate to synchronous speed. However, this speed cannot be attained because at synchronous speed there would be zero slip, hence zero induced rotor voltage, zero rotor current, and zero torque. At no load, that is with no mechanical load connected to the motor shaft, any motor requires a certain torque and rotor current to overcome motor losses. There must be a certain amount of slip in order to obtain sufficient generated secondary voltage to circulate current through the rotor circuit.

In per unit, let base secondary or rotor voltage be the voltage generated in the rotor winding at standstill or unity slip. At any speed of the motor, generated secondary voltage in per unit equals the slip. The amount of current circulating in the rotor winding is equal to rotor voltage divided by rotor impedance. At normal full-load speed, which is only slightly below synchronous speed, rotor frequency is low, and rotor impedance consists essentially of resistance; hence reactance can be neglected. In the per unit system, base current is the secondary current flowing in the rotor winding when the motor develops rated torque. Base secondary resistance is defined as base secondary voltage divided by base secondary current. Slip can then be expressed as

$$s = I_2 R_2 \qquad (8) \blacksquare$$

where I_2 = per unit secondary current
R_2 = per unit secondary resistance

At rated or normal torque, the secondary current is unity. Hence slip is

$$s_{\text{normal}} = R_2 \qquad (9) \blacksquare$$

Since the torque developed by the motor is proportional to the rotor current, torque in per unit is

$$T = I_2 \qquad (10) \blacksquare$$

This means that the slip is proportional to the product of torque and resistance in the rotor circuit.

This basic relation is the same as for d-c shunt motors, which in-

CHARACTERISTICS OF SQUIRREL-CAGE MOTORS

duction motors resemble in many respects. Both types of motors operate with a flux which is essentially independent of load. Speed-torque curves of both motors are straight lines within a certain range of loads, below which the influence of armature reaction can be neglected. Speed is a function of the amount of resistance in the armature circuit (d-c motor) or the rotor circuit (a-c motor). Both motors are capable of developing a definite maximum torque, which is determined by the effect of armature or rotor current, respectively, on the effective flux.

CHARACTERISTICS OF SQUIRREL-CAGE MOTORS

A typical general-purpose squirrel-cage motor is illustrated in Fig. 1-2. There are no external connections to the rotor. After the design of a motor has been established, its speed as a function of load cannot be varied at will. Hence, squirrel-cage motors are used for constant-speed drives, except for polechanging motors, which are discussed later in this chapter.

The starting torque of a squirrel-cage motor depends on the impedance of the rotor winding at standstill, which in turn determines

Fig. 1-2 General-purpose squirrel-cage induction motor.

the magnitude and phase angle of the secondary current. At standstill, rotor frequency equals line frequency and the rotor impedance contains a considerable amount of reactance. The amount of starting torque available depends on the ratio of resistance to reactance since only that component of current which is in phase with the rotor voltage produces torque. The starting torque can be influenced by the choice of material for the rotor conductors, by variations in the shape of the rotor conductors, or by using double cages. In NEMA Motor and Generator Standards, designations have been established for squirrel-cage motors of varying characteristics. These motors are designated as designs A, B, C, D, and F. For each design, certain maximum torques and starting torques have been established, the exact values of which vary with motor frame size, that is to say with speed and horsepower ratings. Most industrial motors are built according to designs B, C, and D. Table 1 lists the maximum and starting torques of motors according to these designs arranged according to speed and horsepower. While individual motors of various manufacturers may differ somewhat from these values, the torque values of Table 1 are minimum values. Speed-torque characteristics of designs B, C, and D are indicated in Fig. 2-2. The shaded areas indicate the range within which the speed-torque curves of a given design of motor fall. Torque values have been standardized for motor ratings up to 200 horsepower. Manufacturers' handbooks should be consulted for data on motors of larger horsepower ratings.

Design B covers a type of motor which is often referred to as "normal starting torque motors." The rotor winding consists of a single cage of copper bars or cast aluminum. Rotor resistance is low; therefore slip and losses are low. The rotor winding is located close to the air gap. As a result, rotor reactance is low, which accounts for high maximum torque. Starting torque is moderate; for large motors, it might even be less than unity.

Motors of design C are also called "high starting torque motors." The rotor winding consists of two squirrel cages. One winding, imbedded deeply in the rotor, has low resistance and high reactance. A second winding, located close to the air gap, has high resistance and low reactance. The combined effect of the two rotor windings produces a lower maximum torque than design B but a considerably higher starting torque. Since slip and running losses are somewhat higher, design C motors are preferred to design B motors only when the increased starting torque can be utilized.

Design D motors have a single squirrel-cage winding similar to de-

Table 1. Maximum and Starting Torques of Various Designs of Squirrel-Cage Motors

Synchronous Speed at 60 Cycles rpm	Horse-power	Per Unit Maximum Torque NEMA Design			Per Unit Starting Torque NEMA Design		
		B	C	D	B	C	D
3600	1½	3.00	—	—	1.75	—	—
	2	2.75	—	—	1.75	—	—
	3	2.75	—	—	1.75	—	—
	5	2.25	—	—	1.50	—	—
	7½	2.15	—	—	1.50	—	—
	10–30	2.00	—	—	1.50	—	—
	40	2.00	—	—	1.35	—	—
	50–60	2.00	—	—	1.25	—	—
	75–100	2.00	—	—	1.10	—	—
	125–200	2.00	—	—	1.00	—	—
1800	1	3.00	—	2.75	2.75	—	2.75
	1½	3.00	—	2.75	2.65	—	2.75
	2–3	2.75	—	2.75	2.50	—	2.75
	5	2.25	2.00	2.75	1.85	2.50	2.75
	7½	2.15	1.90	2.75	1.75	2.50	2.75
	10	2.00	1.90	2.75	1.75	2.50	2.75
	15	2.00	1.90	2.75	1.65	2.25	2.75
	20–25	2.00	1.90	2.75	1.50	2.00	2.75
	30–75	2.00	1.90	2.75	1.50	2.00	2.75
	100	2.00	1.90	2.75	1.25	2.00	2.75
	125–150	2.00	1.90	2.75	1.10	2.00	2.75
	200	2.00	1.90	2.75	1.00	2.00	2.75
1200	¾–1½	2.75	—	2.75	1.75	—	2.75
	2	2.50	—	2.75	1.75	—	2.75
	3	2.50	2.25	2.75	1.75	2.50	2.75
	5	2.25	2.00	2.75	1.60	2.50	2.75
	7½	2.15	1.90	2.75	1.50	2.25	2.75
	10	2.00	1.90	2.75	1.50	2.25	2.75
	15	2.00	1.90	2.75	1.40	2.00	2.75
	20–75	2.00	1.90	2.75	1.35	2.00	2.75
	100–200	2.00	1.90	2.75	1.25	2.00	2.75
900	½–1½	2.50	—	2.75	1.50	—	2.75
	2	2.25	—	2.75	1.50	—	2.75
	3	2.25	2.00	2.75	1.50	2.25	2.75
	5	2.25	2.00	2.75	1.30	2.25	2.75
	7½	2.15	1.90	2.75	1.25	2.00	2.75
	10–200	2.00	1.90	2.75	1.25	2.00	2.75
720	½–1½	2.00	—	—	1.50	—	—
	2	2.00	—	—	1.45	—	—
	3	2.00	—	—	1.35	—	—
	5	2.00	—	—	1.30	—	—
	7½–200	2.00	—	—	1.20	—	—
600	½–200	2.00	—	—	1.15	—	—
514	½–200	2.00	—	—	1.10	—	—
415	½–200	2.00	—	—	1.05	—	—

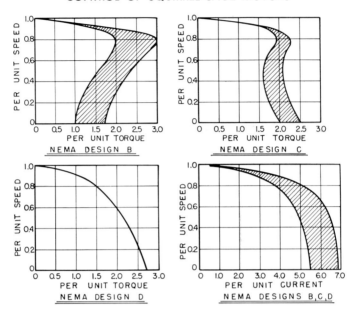

Fig. 2-2 Speed-torque and speed-current curves of general-purpose squirrel-cage motors of designs B, C, and D.

sign B, but the bars are so designed that rotor resistance is increased several times. Design D motors are also known as "high-slip motors." Their slip at full load is considerably higher than the slip of design B and design C motors; therefore the running losses of design D motors are considerably higher. Design D motors possess a high starting torque, which is also the maximum torque. Such motors are used primarily on heavy intermittent-duty drives. They are also used for driving high-inertia loads, the high slip in conjunction with the flywheel effect of the load reducing torque and line-current peaks during load fluctuation.

The highest current which can be drawn from the line by a squirrel-cage motor is its starting current, or "locked-rotor current." Motor manufacturers now list on the motor nameplate a code letter which signifies the locked-rotor current of the motor, expressed as "kva per horsepower" with locked rotor and rated voltage applied to the stator. Code letters, as established by NEMA and listed in the National Electrical Code, are given in Table 2.

The relation between speed and motor current for designs B, C, and

Table 2. Code Letters for Squirrel-Cage Motors

Code Letter	Locked-Rotor kva per hp	Code Letter	Locked-Rotor kva per hp
A	0 –3.15	L	9.0–10.0
B	3.15–3.55	M	10.0–11.2
C	3.55–4.0	N	11.2–12.5
D	4.0 –4.5	P	12.5–14.0
E	4.5 –5.0	R	14.0–16.0
F	5.0 –5.6	S	16.0–18.0
G	5.6 –6.3	T	18.0–20.0
H	6.3 –7.1	U	20.0–22.4
J	7.1 –8.0	V	22.4 and up
K	8.0 –9.0		

D is also indicated in Fig. 2-2. For the motors covered by Table 1, the locked-rotor current varies between 5.5 and 6.75, most motors having a locked-rotor current of 6 or slightly below. It is important that switching devices used for starting squirrel-cage motors are so designed that they are capable of making and breaking locked-rotor current of the motors with which they are used. Underwriters Laboratories, Incorporated, test starters for squirrel-cage motors with currents of six times rated motor current. While this value has worked out quite satisfactorily as an average current, it must be kept in mind that some motors have higher locked-rotor currents. This is particularly true of two-pole motors for which the locked-rotor currents may be as much as ten times full-load current. Some totally-enclosed motors have locked-rotor currents of the order of eight times full-load current. This is the reason why general-purpose industrial starters are capable of making and breaking ten times their rated motor full-load current.

SWITCHING TRANSIENTS

The locked-rotor or starting currents discussed in the preceding paragraphs and indicated in Fig. 2-2 are steady-state symmetrical rms current values such as would be indicated by a switchboard instrument. When motor windings are switched, transient currents may flow which do not have much effect on the starting or steady-state running performance of the motor, but which may have to be considered in the selection of control equipment and accessories.

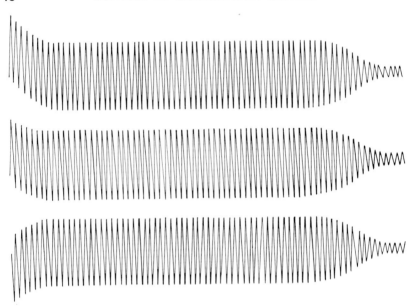

Fig. 3-2 Starting currents in three phases of squirrel-cage motor.

When a squirrel-cage motor is connected to the line at standstill, the initial starting current contains an offset; that is to say, the a-c current wave includes a d-c component, the magnitude of which depends on the point on the voltage wave at which the connection to the motor winding is established, and the decrement of which depends on the time constant of the motor windings. The time constant of small and medium-size motors is of the order of a few cycles; for large motors, it may be as high as 1 second. The actual starting current of a three-phase induction motor is indicated in Fig. 3-2. The current traces are shown for the three phases. The symmetrical starting current is assumed to be six times the running current. The current at the top is drawn under the assumption that maximum offset occurs, and the peak current of the first cycle rises to nearly twice the peak value of the steady-state symmetrical current. The other two phases have a lesser offset. When squirrel-cage motors are connected to circuits which are protected by circuit breakers with instantaneous current trips, the breaker setting must take into consideration that one phase may see an initial current peak nearly twice the symmetrical starting current. Since circuit-breaker trips or instantaneous relays

SWITCHING TRANSIENTS

may respond within one cycle, instantaneous setting of protective devices should be about twelve to fifteen times motor full-load current.

The traces of Fig. 3-2 indicate that the current transient exists only for a rather short portion of the total starting time. The starting current does not change appreciably during the major portion of the starting period, but drops fairly rapidly as normal running speed is approached. After acceleration is complete, normal running current flows. Another switching transient occurs when a motor is disconnected from the line and reconnected after a very short time interval. The flux of a motor that has been disconnected from the line, does not collapse immediately, but decays exponentially. The time constant of decay of the entrapped flux varies considerably. It may be only a few cycles for small motors. For many medium-size motors, it is of the order of 1 second. For large high-speed motors, it may even be several seconds.

If a motor is disconnected from the line and then reconnected before its entrapped flux has decayed to zero, the entrapped flux is no longer in synchronism with the flux pattern which would correspond to the line voltage. The shift in flux pattern results in a high transient current peak of short duration, which may cause excessive torques on the motor shaft, high stresses in the motor windings, particularly the end turns, and excessive voltage dips in the power system. This transient current peak is illustrated in a general manner in Fig. 4-2. The motor current is represented as the rms current of one motor phase, either a running current or a starting current. When the motor is disconnected, current flow ceases immediately; but the motor continues to run, and its flux is entrapped. When the motor is reconnected to the line a short time later, a high current peak results which decays rapidly

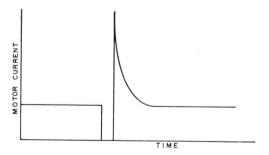

Fig. 4-2 Current peak resulting from reconnecting motor after complete interruption.

to the normal motor current. The magnitude of the current peak depends on many variables in the motor design, on the time interval between disconnection and reconnection, and on the phase angle at which the motor is reconnected. It is difficult to predict a definite magnitude for any particular motor, but current peaks considerably in excess of normal starting current have to be expected.

When a small or medium-size motor is disconnected and immediately reconnected to the line by control action or by a short-time interruption in the power supply, the transient current peak is of no significance. In practically all cases, general-purpose motors are reconnected immediately to the line with no apparent harm and no significant reduction in service life due to mechanical and electrical stresses. End turns are short and comparatively strong; hence motors are capable of absorbing the extra stresses. Large motors, particularly high-speed motors used for power-station auxiliaries, may be damaged by the extra stresses resulting from these switching transients. Starters for large motors may include provision to prevent their being re-energized until a sufficient period of time has elapsed to permit the extrapped flux to decay to a very low value.

Similar transient current peaks occur when motor connections are switched during starting steps or when motor connections are changed. Controllers which disconnect the motor from the line and then re-energize it with changed connections are called controllers with "open-circuit transition." With such controllers high switching current transients can be expected when transferring from one step to the next one. If these transient current peaks are objectionable, controllers with "closed-circuit transition" are used. With such controllers, the motor is not disconnected from the line between steps, so that the motor flux pattern remains in synchronism with the line voltage pattern.

DUAL-VOLTAGE MOTORS

Standard voltages for industrial type a-c motors are 110, 220, 440, and 550 volts. Very few polyphase industrial motors are connected to 110-volt systems. Distribution at 110 volts is generally for single-phase lighting, domestic appliance, and commercial loads. Very few 110-volt polyphase distribution networks exist today in industrial plants, and none are being installed. Likewise, there are few industrial systems in existence at the 550-volt level, so that the market for 110-volt and 550-volt motors is limited.

The bulk of industrial polyphase motors operate at 220 or 440 volts.

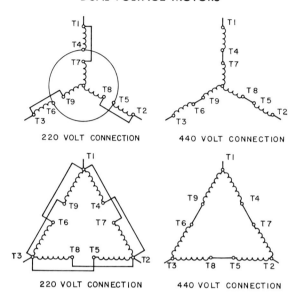

Fig. 5-2 Connection of 220/440-volt reconnectable induction motors with wye- or delta-connected stator winding.

Motors operating at 220 volts are used primarily on 240/120- or 208/120-volt three-phase, four-wire distribution networks as are found generally in downtown and suburban distribution systems. Such systems serve combination power and lighting loads, motors being connected to the three-phase 220- or 208-volt lines, whereas lighting and appliances are connected to the 120-volt single-phase lines. Small industrial loads are generally connected to such systems. Large industrial plants generally have 480-volt three-phase power-distribution systems, which serve motor loads exclusively. Thus the majority of general-purpose industrial motors are used on 440-volt service.

Standard general-purpose small and medium-size three-phase motors are built as dual-voltage motors; that is to say, they can operate on either 220 or 440 volts. This standardization improves manufacture, reduces costs, and simplifies the stocking of standard motors. Such motors are built with two legs in each phase winding, and nine leads are brought out for connection by the user for either 220- or 440-volt operation. Standard connections are indicated in Fig. 5-2. T1, T2, and T3 are the terminals which are connected to the line. Both wye and delta connection are indicated.

FULL-VOLTAGE STARTERS

All general-purpose squirrel-cage motors which are built today can be started by applying full voltage across the stator terminals. Such starters, which start the motor in a single step are called full-voltage starters. They are also called "across-the-line starters" because the motor is immediately connected across the line. Different methods of starting, in which the motor is connected to full line voltage in several steps, are used for reasons of the driven machinery or because of limitations on the power system to which the motor is connected.

Full-voltage starters consist essentially of a line contactor and a pair of overload relays. Figure 6-2A shows a size 2 starter, which is representative of the smaller starters employing lift-type contactors. The overload relays are mounted on the side of the magnet. Figure 6-2B shows a size 5 starter, using a shaft-type contactor with separately mounted overload relays. On still larger starters, current transformers may be used, and the overload relays are connected in the secondary circuit of the current transformers. The current or horsepower rating of the starter is determined by the capabilities of the line contactor. Starters for specific motor loads are listed in the tables given in Chapter 2 of Part I. The overload relays serve to disconnect the motor from the line in case it is overloaded. The function of these relays will be discussed in greater detail in Chapter 6.

NEMA has established standard connections for full-voltage starters, as indicated in Fig. 7-2. Standard starters are built with two overload relays connected in lines $L1$ and $L3$. The control circuit is connected to lines $L1$ and $L2$. If one line of the power system is grounded, $L2$ is the grounded line. Figure 7-2A shows the most commonly used control circuit with momentary start-stop pushbutton and with the control circuit at line voltage. This control circuit is sometimes called a "three-point control circuit," because three control lines marked 1, 2, and 3 are carried from the starter to the pushbutton station. Momentarily pressing the start button energizes the line contactor M, which seals itself in through a normally open interlock, so that the start button may be released without dropping out the line contactor. Pressing the stop button or tripping one of the overload relays drops out line contactor M, causing the motor to become disconnected from the line and to shut down. The start button has to be pressed again to restart the motor.

A similar control circuit, using a spring-return master switch in place of the pushbutton station, is shown in Fig. 7-2B. Turning the master

FULL-VOLTAGE STARTERS 21

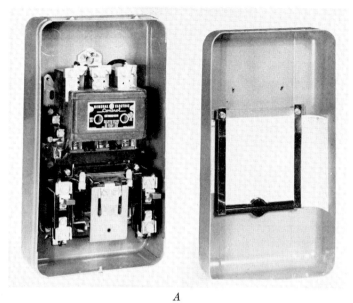

A

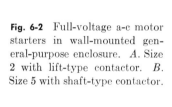

Fig. 6-2 Full-voltage a-c motor starters in wall-mounted general-purpose enclosure. *A*. Size 2 with lift-type contactor. *B*. Size 5 with shaft-type contactor.

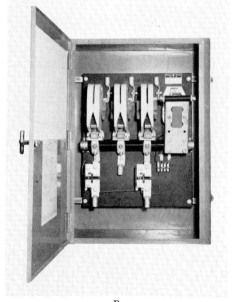

B

22 CONTROL OF SQUIRREL-CAGE MOTORS

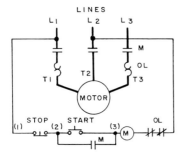

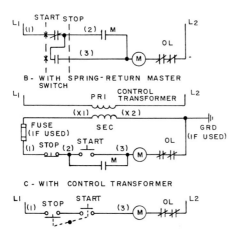

Fig. 7-2 Standard connections of full-voltage a-c starters with various types of control circuits. M = line contactor; OL = overload relays.

switch in the start position energizes line contactor M. Releasing the master-switch handle causes the master switch to return to the neutral position. Momentarily turning the master switch in the stop position drops out the line contactor and shuts down the motor. This circuit also requires three lines from the starter to the master switch.

Some users, particularly machine-tool builders and high-production manufacturing plants, prescribe that the control circuit carry no more than 110 volts, as a safety measure. This requirement is included in the published electrical standards of the National Machine Tool Builders Association and the Joint Industry Conference. Since 440 volts is the preferred voltage for motor circuits, a control transformer has to be used. Figure 7-2C shows standard control connections with a control transformer. The primary winding of the transformer is connected to lines $L1$ and $L2$. The control circuit is connected to the secondary

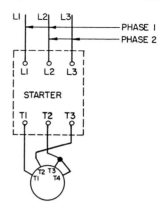

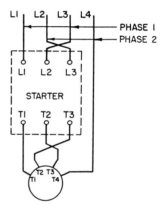

Fig. 8-2 Connections of three-phase full-voltage starters when used with two-phase motors.

A- WITH TWO-PHASE THREE-WIRE LINES

B- WITH TWO-PHASE FOUR-WIRE LINES

winding $X1$–$X2$. If protection of the control circuit is specified by the user, a fuse or single-pole circuit breaker is connected in the $X1$ line. If the user desires to ground the control circuit, the ground is applied to the $X2$ line.

Starters are sometimes used with maintained-type control-circuit devices, as indicated in Fig. 7-2D. This particular circuit is shown with a maintained-type pushbutton station. Pressing the start button closes the control circuit, and the pushbutton stays in the closed position. Pressing the stop button opens the control circuit and disconnects the motor. Since this type of circuit requires only two control leads between the starter and the control-circuit device, it is sometimes called "two-point control." This circuit is most often used when the starter

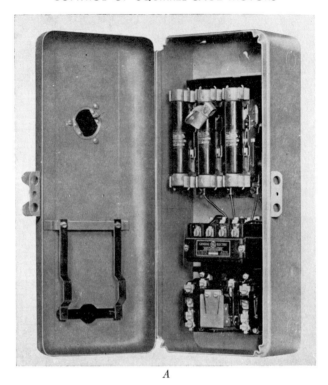

Fig. 9-2 A-c low-voltage combination starters. *A*. With fused motor circuit switch. *B*. With molded-case air circuit breaker.

is under the control of an automatic control-circuit device, such as a pressure switch, vacuum switch, float switch, and the like.

Standard across-the-line starters do not contain control-circuit fuses. These fuses are often included in special controllers, particularly in complex multimotor equipments. If control-circuit fuses are desired by the user, they have to be specified.

Occasionally, standard three-phase starters are used to control two-phase motors. Since the number of two-phase motors installed annually is comparatively small, it would be uneconomical for control manufacturers to build and stock special starters for these motors. Three-phase starters can be used on two-phase motors, and they provide adequate protection. It is important to connect the lines and the motor terminals to the correct starter terminals, so as to place the overload relays in the two phases. Figure 8-2*A* shows the proper connec-

COMBINATION STARTERS 25

B

Fig. 9-2 (Continued)

tions for a motor and starter to a three-wire system. Care must be taken that $L2$ is the common phase line. Figure 8-2B shows correct connections of a motor and starter to a four-wire system. When a three-phase starter is used, line $L4$ is carried directly to the motor and is not switched by the starter. This means that, with the starter dropped out, motor terminal $T4$ is hot. Where this feature is objectionable, a starter with a four-pole contactor must be used.

COMBINATION STARTERS

Standard starters include overload protection, but they do not include disconnecting means or short-circuit protection. The National Electrical Code requires that each motor branch circuit include disconnecting means and short-circuit protection, but it does not require these features to be included as part of the controllers. They may be provided as separate components. The Industrial Control Industry offers so-called "combination starters," which include a standard starter with a fused motor circuit switch or a molded-case air circuit breaker mounted in a common enclosure. Figure 9-2 shows such combination starters. How short-circuit protection and motor overload protection are coordinated with the characteristics of the power system will be discussed in greater detail in Chapter 7.

Connections of combination starters do not differ basically from connections of standard full-voltage starters. Combination starter connections with momentary-contact, start-stop pushbutton stations are shown in Fig. 10-2. Control circuits are connected to phases 1 and 2. Control circuits are connected to the load side of the disconnecting devices so that, when the motor circuit switch or the air circuit breaker is opened, power as well as control circuits are disconnected and no terminals on the starter are hot.

UNDERVOLTAGE PROTECTION

Where a starter is controlled by a maintained-contact control-circuit device, as shown in Fig. 7-2D, the line contactor will drop out and the motor will stop when the power supply is interrupted. When voltage reappears in the power circuit, the line contactor recloses immediately and restarts the motor. This behavior is called "undervoltage release," and is suitable for use with machinery which starts and stops automatically without the aid of an attendant. For machinery that is started and stopped manually by an operator, however, undervoltage release may constitute a safety hazard. When the drive is automatically restarted upon reappearance of voltage after a power outage, the operator may be injured by inadvertent restarting of the machine.

Momentary-contact, start-stop pushbutton stations and spring-return master switches, as shown in Fig. 7-2A, B, C, provide "undervoltage protection." When the drive is stopped by loss of power supply voltage, the motor is not restarted immediately upon reappearance of voltage. The operator has to press the start button or turn the master switch in the start position before the line contactor is energized again. Thus undervoltage protection is an automatic feature of momentary-contact pushbuttons and spring-return master switches, which means that a machine operator is protected against possible injury from the inadvertent restarting of his machine. For machinery operated by an attendant through maintained-contact control-circuit devices or master

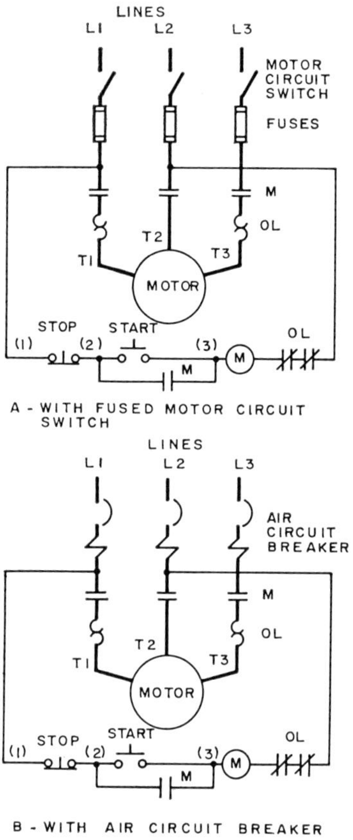

UNDERVOLTAGE PROTECTION

Fig. 10-2 Standard connections of full-voltage a-c combination starters. M = line contactor; OL = overload relays.

A - WITH FUSED MOTOR CIRCUIT SWITCH

B - WITH AIR CIRCUIT BREAKER

switches, it may be necessary to provide an undervoltage relay, so that the operator is forced to return his master switch to the start position to avoid inadvertent automatic restarting.

Since a-c contactors have a short drop-out time, a system disturbance resulting in loss of voltage for even a very short duration would shut down magnetically controlled motors. It may be desired, particularly in automated plants, to ride through short interruptions of power. This can be accomplished by time-delay undervoltage protection. A time-delay undervoltage pushbutton station, as shown in Fig. 11-2, contains a standard start-stop pushbutton and a time-delay undervoltage relay and accessory components mounted in a common enclosure. The undervoltage protective circuit is detailed in Fig. 12-2. A half-wave semiconductor rectifier supplies d-c power for the coil

28 CONTROL OF SQUIRREL-CAGE MOTORS

circuit of auxiliary control relay CR. When the start button is pressed, relay CR is energized. A contact on this relay in turn energizes line contactor M. A capacitor CAP is connected in parallel with the coil circuit of relay CR. When a-c voltage disappears, line contactor M drops out immediately and opens the motor circuit. However, control relay CR does not drop out right away, since its coil circuit remains energized because of the discharge of capacitor CAP through the relay coil. Thus CR drops out with a time delay. If voltage reappears before relay CR has dropped out, line contactor M is reclosed immediately without the need of pushing the start button. However, if the loss of a-c voltage lasts longer than the drop-out time of relay CR, M does not reclose automatically, and the start button has to be pressed to restart the motor.

Careful judgment must be exercised in considering the use of time-delay undervoltage protection. The question should be asked whether the driven machinery can be restarted automatically without damage to it. As an example, when a vertical deep-well pump is disconnected from the line, the column of water above the impeller will drive the pump in the reverse direction. Reenergizing the motor while the pump

Fig. 11-2 Time-delay undervoltage pushbutton station.

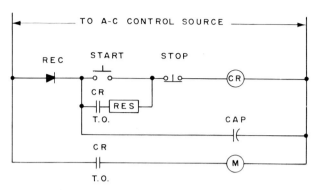

Fig. 12-2 Time-delay undervoltage protective circuit. M = line contactor; CR = control relay.

spins backward may cause damage to the pump shaft. Small and medium-size general-purpose motors are generally not harmed by the switching transient which may accompany automatic restarting with the flux still entrapped in the motor. However, where large motors are involved, a check should be made to determine whether the switching transient current peak might not cause damage to the motor.

It should also be ascertained what effect the simultaneous starting of a multiplicity of motors may have on the stability of the power system. Simultaneous starting of a large number of motors may draw such a high starting current from the power system that the resultant voltage drop may prevent the motors from accelerating their load. Hence, time-delay undervoltage protection should be used only for a limited number of motors connected to a power system.

JOGGING

"Jogging" or "inching" is defined by NEMA as the quickly repeated closure of the circuit to start a motor from rest for the purpose of accomplishing small movements of the driven machine. This means that the motor runs only for short intervals, as controlled by the operator. Jogging is frequently used on production machinery, for instance to bring the tool of a machine tool into alignment with the work or to thread cloth into a printing machine.

Several jogging control circuits are shown in Fig. 13-2. Circuits A and B can be used with standard starters; circuit C requires a special starter. The simplest jogging control is obtained by a single momentary-contact pushbutton, as shown in sketch A. Contactor M is

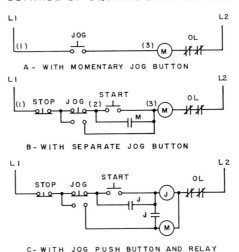

Fig. 13-2 Jogging control circuits. M = line contactor; OL = overload relays; J = jogging relay.

energized, and the motor turns only as long as the pushbutton is held depressed. This circuit is only used for auxiliary drives which are used for occasional position adjustments but which do not run for prolonged periods of time. In most instances, it is desired to jog the machine and then to start it for a long-time run.

In Fig. 13-2B, a jog button is provided in addition to the regular start and stop buttons. When the jog button is pressed, the normally closed back contact interrupts the sealing circuit through the interlock on M. Therefore M drops out as soon as the jog button is released. This method has a serious disadvantage. If the jog button is released suddenly, its normally closed back contact may close before contactor M has dropped out, thereby reestablishing the sealing circuit and restarting the motor. If the operator is intent on inching his work he may be hurt by the unexpected starting of the machine. To overcome this disadvantage, jog pushbuttons can be equipped with a mechanical latch which prevents the normally closed back contact from reclosing as long as the latch is turned.

A safer method is to use a special starter for jogging, as illustrated in Fig. 13-2C. A jog relay J is added to the starter. When the start button is pressed, relay J is energized, which in turn energizes contactor M and provides the sealing circuit. When the jog button is pressed, contactor M is energized directly, and relay J is not in the circuit. With this connection, relay J is positively prevented from closing when the normally closed back contact of the jog button recloses, and there is no danger of contactor M staying energized. Although the control circuit according to Fig. 13-2C involves the expense of an additional relay, it offers a higher degree of safety to the operator.

When small- and medium-size motors are jogged, the time interval between jogs is generally long compared with the time required for the entrapped flux of the motor to decay after the motor has been deenergized. Switching transient current peaks upon repeated reenergizing the motor are generally not a problem. When motors and controllers are jogged frequently, that is to say more than five times per minute, the frequent jogging duty has to be considered in the selection of motors and starters because the frequent incidence of starting current inrushes presents a thermal problem. Starters as well as motors selected for normal duty would overheat when applied to frequent jogging duty.

When large motors are to be jogged, the time required for the entrapped flux to decay may be longer than the time interval between jogs. If that is the case, high switching transient current peaks may result when the motor is jogged with short time intervals between jogs. When the jogging of large motors is considered, it should be checked with the motor manufacturer whether the motor is suitable for jogging duty.

INDEXING

"Indexing" or "cycling" means that the motor is started, runs for a predetermined distance of the driven machine, and then stops automatically. This cycle is repeated as often as the operator desires. An example is a rotary car dumper which starts, rotates through a complete revolution, and then stops automatically. The point of stopping, which is also the starting point for the next cycle, is called the "index point." Indexing can be controlled by limit switches. Several circuits which accomplish this function are shown in Fig. 14-2.

Circuit A shows limit switch LS actuated by the driven machinery. In the index position, LS is open. When the operator presses the start button, contactor M closes and the motor starts. The operator must keep the start button depressed for a certain length of machine travel until the limit switch recloses. When the pushbutton is then released, contactor M is held closed through LS and its own holding interlock. When the index position is reached, LS opens, drops out M, and stops the motor. This circuit has the advantage of simplicity. Its dis-

CONTROL OF SQUIRREL-CAGE MOTORS

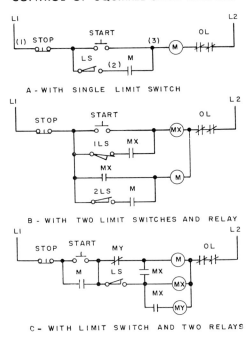

Fig. 14-2 Control circuits for indexing or cycling by limit switches. M = line contactor; MX, MY = control relays; LS, $1LS$, $2LS$ = limit switches.

advantage is that, when a new cycle is started, the start button must be held depressed for a certain length of time, according to the judgment of the operator. On slow-moving drives, considerable time may elapse before LS recloses, during which the operator must hold the start button closed.

This disadvantage is avoided by circuit B. Two limit switches are provided, $1LS$ and $2LS$. In the index position, $1LS$ is closed and $2LS$ is open. When the start button is pressed, relay MX is energized, seals itself in through $1LS$, and energizes contactor M. Shortly after the drive has started, limit switch $2LS$ recloses and establishes a sealing circuit for M. A short distance beyond this point, limit switch $1LS$ opens and causes relay MX to drop out. A short distance before the index point is reached again, $1LS$ recloses, but no change takes place in the control circuit. When the index point is reached, $2LS$ opens and drops out M, thus stopping the motor. Momentarily pressing the start button again restarts the cycle.

HIGH-VOLTAGE STARTERS

Circuit C eliminates one of the limit switches, but another relay is added. One limit switch LS is used, which opens at the index point. Pressing the start button energizes contactor M through the normally closed contact of relay MY. Contactor M seals itself in. After the drive has traveled a short distance, LS recloses and energizes relay MX, which seals in contactor M through LS. Relay MX in turn energizes relay MY, which opens its contact and thus places M under the control of LS. At the index point, LS opens, drops out MX and M, and stops the motor. Relay MY drops out and recloses its contact. Pressing the start button momentarily restarts the cycle. It would be possible to eliminate relay MY by using a normally closed contact on MX, so adjusted that it opens after the normally open contacts on MX close. However, since such an adjustment is difficult to maintain, it is preferable to use a separate relay MY.

With any one of the control circuits shown in Fig. 14-2, it is possible to intercept the cycle and stop the motor by pressing the stop button. The cycle can then be completed by pressing the start button again.

HIGH-VOLTAGE STARTERS

Starters for high-voltage motors employ control circuits similar to those used on low-voltage starters. Some special features have to be included because of the nature of the high-voltage power circuits. Connections of a typical starter are indicated in Fig. 15-2. Since inadvertent contact with high-voltage circuits is much more lethal than contact with low-voltage circuits, most high-voltage starters built today include an isolating switch which serves to disconnect the starter from the line and gives added assurance to an electrician working on the starter that the high-voltage circuits cannot be energized by mistake. The isolating switch is capable of interrupting the control power, but it is generally not a motor circuit switch capable of interrupting motor running current. Current transformers are needed for the overload relays, since it would be impractical to design such relays for direct connection to high voltage. Likewise, the control circuits are fed from a control transformer. To safeguard against short circuits in the control circuits and against internal faults in the primary winding of the control transformer, both primary and secondary circuits of the control transformer are fused. One side of the control circuit is grounded to prevent raising the control-circuit potential to high voltage by accidental contact between low-voltage and high-voltage conductors. A safety interlock in the control circuit in-

CONTROL OF SQUIRREL-CAGE MOTORS

sures that the isolating switch is closed and that doors providing access to the high-voltage sections of the controller are closed. This is very important, since the isolating switch must not be closed upon a load, and accidental contact of operating personnel with energized high-voltage circuit components must be avoided. The safety interlock which assures that this important safety requirement has been fulfilled is often a door interlock which is closed by the movement of the door or by a handle actuating the door latch.

High-voltage motors are generally fairly large motors, and they are often used on drives on which automatic restarting after a dip or short-time outage of system voltage is desirable. The control circuit as shown in Fig. 15-2 includes an undervoltage relay UV, which can be set up for either instantaneous or time-delay undervoltage protection by a simple shift in connections on the controller terminal board. The coil of relay UV is connected to a rectifier circuit, and time delay drop-out is obtained by capacitor discharge in the same manner as previously described for Fig. 12-2. Upon loss of control voltage, the coil of relay UV is kept energized for a period of time by the discharge of the capacitor. However, when the stop button is pressed,

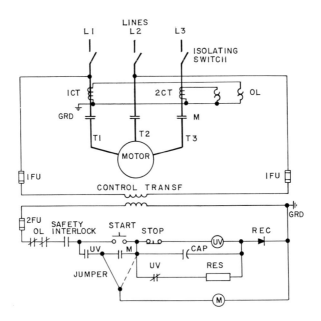

Fig. 15-2 High-voltage full-voltage starter. M = line contactor; UV = undervoltage relay; OL = overload relays.

REVERSING

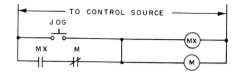

Fig. 16-2 Antikiss circuit for jogging high-voltage starters. $M =$ line contactor; $MX =$ auxiliary control relay.

the UV coil circuit is interrupted, the relay drops out instantly, and the capacitor is rapidly discharged through a resistor.

With the jumper connected as indicated by the solid line, loss of voltage causes the line contactor M to drop out. However, since UV remains closed for a time, M is reenergized if voltage returns before UV has dropped out. When the jumper is shifted to the position indicated by the broken line, contactor M is not reenergized automatically when voltage returns, but the start button has to be pressed to restart the motor.

Because of their size, high-voltage motors are rarely jogged. There are some applications, however, particularly in the rubber industry, which require frequent jogging of the motors. If the operator tries to jog rapidly and releases the jog button before the contactor has fully closed, it is possible to let the contacts "kiss" and to deenergize the magnet before the contacts have gone through their rolling and wiping motion. Contacts are likely to weld closed when they just touch and "make" the inrush current of the motor. This welding can be prevented by the use of an "antikiss" circuit, as illustrated schematically in Fig. 16-2. When the jog pushbutton is pressed, an auxiliary relay MX is energized simultaneously with line contactor M. This relay closes much faster than M and, through its normally open interlock, keeps the coil circuit of M energized when the jog button is released immediately. When M has fully closed, that is to say when its contacts have gone through their rolling and wiping motion, the normally closed interlock on M opens. If meanwhile the jog button is released, M will drop out. If the jog button is still depressed, M will stay closed. But irrespective of how fast the jog button is released, M always closes completely once its coil is energized.

REVERSING

Polyphase a-c motors are reversed by interchanging two stator leads. A pair of reversing contactors is necessary to accomplish motor reversing. When one contactor is closed, the motor is connected to the line with phase rotation for one direction of rotation. When the other contactor is closed, the motor is connected to the line with reversed

Fig. 17-2 Full-voltage reversing starter.

phase rotation. A typical reversing starter for a squirrel-cage motor, consisting of two contactors in a common enclosure, is shown in Fig. 17-2. Simultaneous closing of both contactors must be prevented, since this would result in short-circuiting the line. For this reason all reversing starters are equipped with mechanical interlocks which permit only one contactor to be closed at one time. The mechanical interlock shown in Fig. 17-2 consists of two cams which are so shaped that, when one of the two contactors is energized, its cam pivots and by mechanical interference prevents the cam of the other contactor from turning, thus preventing that contactor from being closed.

NEMA has standardized connections for general-purpose reversing starters, as shown in Fig. 18-2. To reverse the motor, phases 1 and 3 are interchanged. In addition to mechanical interlocking between the

REVERSING

two reversing contactors, electrical interlocking may be used. If the pushbuttons have back contacts, they may be connected as shown in the diagram. This cross connection of pushbuttons prevents simultaneous energizing of the two reversing contactor coils in the event that the forward and reverse pushbuttons are pressed simultaneously. Electrical interlocking can also be accomplished by normally closed interlocks on the contactors. The use of electrical interlocks is not mandatory, and many reversing starters with interference-type mechanical interlocks are built without electrical interlocks.

Reversing starters may be used on drives which are required to stop at the ends of the forward and reverse travel. Limit switches are then used to stop the motor automatically at these points. In Fig. 18-2, limit switches are shown connected in the reversing contactor coil circuits.

When high-voltage reversing starters are equipped with oil-immersed contactors, control circuits are the same as those for low-voltage re-

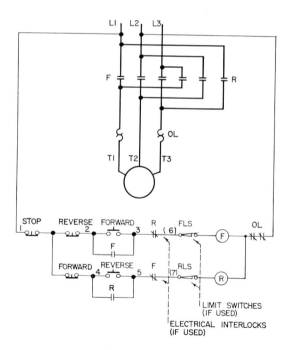

Fig. 18-2 Standard connections of full-voltage reversing starter. F = forward contactor; R = reverse contactor; OL = overload relays; FLS = forward limit switch; RLS = reverse limit switch.

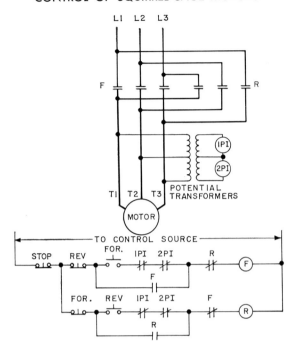

Fig. 19-2 Potential interlocking circuit for high-voltage air-break reversing contactors. F = forward contactor; R = reverse contactor; $1PI$, $2PI$ = potential interlocking relays.

versing starters. However, if high-voltage motors are reversed by air-break contactors, electrical and mechanical interlocking is no definite insurance against possible short-circuiting of the line. On high-voltage air-break contactors, arc duration is longer than on oil-immersed contactors and low-voltage contactors. The arcs across the stationary and movable contacts may not be extinguished when the contactors are dropped out. If the reverse contactor should close before the arcs on the forward contactor are extinguished, the line would be short-circuited. This faulty operation can be avoided by potential interlocking in addition to the conventional electrical and mechanical interlocks.

Figure 19-2 describes the essentials of potential interlocking. While this is not a complete control circuit, the intent is to illustrate the functioning of a potential interlocking scheme which can be worked into any complete reversing controller. Potential transformers are connected to the load side of the reversing contactors, and potential interlocking relays $1PI$ and $2PI$ are connected to the secondary side of the potential transformers. Each relay has two normally closed contacts, which are connected in the coil circuits of the reversing contactors. As long as either reversing contactor is closed, or as long as any arcs exist across the contacts of either contactor, one or both of the potential interlocking relays are energized and their contacts are open. After one reversing contactor has dropped out, the other contactor cannot close until after $1PI$ and $2PI$ have both dropped out, which means that all arcs have been extinguished. The reversing contactors seal themselves in through normally open interlocks which bypass the potential interlocking-relay contacts once the motors have been started.

REGENERATIVE BRAKING AND PLUGGING

In the preceding paragraphs, squirrel-cage motor operation has been considered only for the condition that the motor drives a positive load. Referring to Fig. 1-1, this corresponds to quadrant 1 in the case of a nonreversing drive, or to quadrants 1 and 3 in the case of a reversing drive. Operation of the motor is also possible in quadrants 2 and 4, which represent the condition that the motor is driven by the load. The motor is capable of developing a braking torque which opposes load torque. Speed-torque curves of a typical squirrel-cage motor in all four quadrants are plotted in Fig. 20-2.

If a squirrel-cage motor is driven by a load in the same direction as

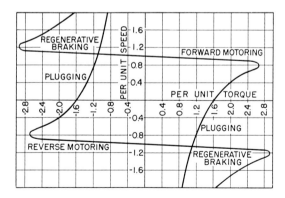

Fig. 20-2 Typical regenerative braking and plugging curves of a squirrel-cage induction motor.

the rotation of its flux, its speed rises above synchronous speed. The motor acts as an induction generator, taking magnetizing current from the line and absorbing mechanical power through its shaft. Electric power is fed back into the power system. This condition exists when the motor restrains a load hanging on the hook of a hoist or when the load overhauls an inclined conveyor. In both cases, the load is under the influence of gravity and drives the motor. Its slip becomes negative, the motor develops a braking torque, and the speed-torque curve is similar to the motoring curve, except that it is located in the dynamic braking quadrants 2 or 4. Maximum torque is somewhat higher than for motoring operation, owing to the influence of stator resistance. The increase in maximum torque may vary from a small percentage for large, low-resistance motors to as high as 50 per cent for small, high-resistance motors. The motor restrains the load with little rise in speed above synchronous, as long as the load torque does not exceed maximum torque of the motor. If maximum torque is exceeded, the motor becomes unstable and runs away. As the motor draws its excitation from the line, no dynamic braking action is possible when the connection between the motor stator and the a-c power line is interrupted. Regenerative braking at speeds below synchronous is not possible. Retardation of the motor to standstill must be accomplished by other means.

A squirrel-cage motor may be retarded quickly by plugging. This means that, while the motor is running at full speed in one direction, its stator connections are reversed. The motor runs in a direction opposite to the rotation of its flux, and slip is greater than unity. Speed-torque curves for plugging are also indicated in Fig. 20-2. They are also located in the braking quadrants 2 and 4. The plugging torque available at full speed varies approximately between the limits of one half and full motor starting torque, the exact amount depending on the resistance of the motor winding. The higher the resistance, the higher is the ratio of plugging torque to starting torque. Plugging can be used advantageously to brake a motor driving an inertia load to a quick stop, because the plugging torque opposes the overhauling inertia torque of the load. Plugging is not suitable for restraining a load under the influence of gravity, such as a load on a crane hook, because plugging torque decreases with increasing speed.

During plugging operation, motor current is slightly higher than starting current. The energy transmitted from the load to the motor is not fed back into the line, but is dissipated within the motor as

PLUGGING CONTROL

losses. When a squirrel-cage motor is applied to a drive which is plugged frequently, care must be taken to select a motor having sufficient thermal capacity to dissipate the losses occurring during plugging operations. Contactors used for plugging a motor frequently must also be selected from a rating table which has been established for frequent plug-stop duty.

PLUGGING CONTROL

Any reversing starter can be used for plugging the motor manually, as shown in Fig. 18-2. Suppose the motor runs in the forward direction. Pressing the reverse button drops out the forward contactor and closes the reverse contactor. The motor is then plugged. When the motor has come to a standstill, the stop button may be pressed, bringing the motor to a plug stop. The motor may be permitted to accelerate in the reverse direction. Plugging is then used for rapid reversal.

Plugging control is often employed to bring the motor to a plug stop automatically, without accelerating in the reverse direction. Connections for automatic plug-stop control are shown in Fig. 21-2. A plugging switch is coupled to the motor shaft. The switch closes its contacts when the motor rotates in either direction; at standstill, its contacts are open. This switch is used as an indication that the motor has come practically to a stop. When there are occasions requiring an operator to turn the motor by hand, as when adjusting the work on a machine tool, the plugging switch is equipped with a lock-out coil, which prevents the switch contacts from being closed unless the motor is energized from the line. The lock-out coil prevents accidental energizing of the motor which might inadvertently start the machine and injure the operator.

Plug-stop control for a nonreversing drive is shown in Fig. 21-2A. When the start button is pressed, contactor F closes and connects the motor to the line in the forward direction. The coil circuit of contactor R is interrupted through a normally closed interlock on F. As the motor starts rotating, the plugging switch closes its forward contact. When the stop button is pressed, contactor F is dropped out and contactor R closes, since its coil is energized through the plugging switch. Motor stator connections are reversed, and the motor develops a countertorque opposed to the load torque, which results in rapid retardation. When the motor speed has dropped to nearly zero, the plugging switch opens and contactor R is dropped out. The motor is disconnected from the line and coasts to a complete stop.

Plug-stop connections for a reversing drive are shown in Fig. 21-2B.

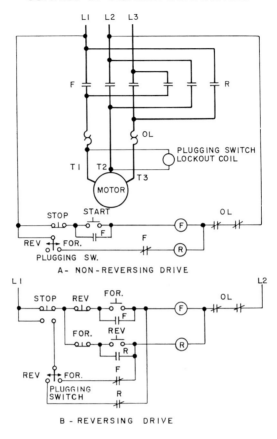

Fig. 21-2 Plug-stop control connections for squirrel-cage motor. F = forward contactor; R = reverse contactor; OL = overload relays.

The motor is started in either direction of rotation by pressing the forward or reverse pushbutton in the conventional manner. As the motor accelerates, the plugging switch closes either its forward or its reverse contact. The stop button must be of the heavy-duty type; that is to say, it has a normally closed circuit to give the conventional stop function. In addition, the normally open circuit is used to energize the plugging control circuits through the plugging switch. To obtain automatic plug stop, it is necessary to press the stop button all the way down and hold it depressed. When the motor has been running forward, the R contactor is energized through the plugging switch. If the motor has been running in the reverse direction, contactor F is energized through the plugging switch. The motor is disconnected automatically from the line when the plugging switch opens its contact near zero speed. If the stop button is pressed but released again before the motor has been plugged to a stop, the plugging connection is broken and the motor coasts to a stop.

DIRECT-CURRENT DYNAMIC BRAKING

Another method of bringing a squirrel-cage motor to a quick stop is the use of direct-current dynamic braking. If the stator of an induction motor is excited with direct current, a d-c flux is set up in the motor, which is stationary in space. When the motor is driven by a load, the rotor conductors intersect the d-c flux. An emf is generated in the rotor conductors, causing current to flow in the rotor. This current, in conjunction with the d-c flux, develops torque which is opposed to the torque driving the motor.

Any two of the stator leads are connected to the d-c source, the third lead being idle. The torque that can be obtained with a given amount of d-c excitation varies with the specific design of the motor, its number of poles, and its rotor impedance. For specific applications, braking data must be obtained from the motor designer. Figure 22-2 shows typical dynamic braking speed-torque curves. Braking torque varies with the amount of d-c excitation. Higher average braking torque over the speed range can be obtained than with plugging. Direct-current dynamic braking is suitable for applications in which it is desired to retard a load rapidly. The braking torque reaches a peak at low speed. When the motor has been brought to a stop, the torque

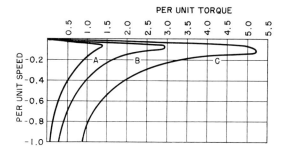

Fig. 22-2 D-c dynamic braking curves of a typical squirrel-cage induction motor for various amounts of d-c excitation. *A*. 1.33 a-c full-load current. *B*. 2.66 a-c full-load current. *C*. 4.0 a-c full-load current.

44 CONTROL OF SQUIRREL-CAGE MOTORS

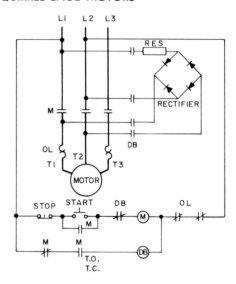

Fig. 23-2 D-c dynamic braking control for squirrel-cage induction motor. M = line contactor; DB = dynamic braking contactor; OL = overload relays.

becomes zero. The motor will not start to rotate in the opposite direction, which is an important advantage over the plugging method.

Connections of a squirrel-cage motor starter with d-c dynamic braking are shown in Fig. 23-2. For normal running, line contactor M is closed and dynamic braking contactor DB is open. Line contactor M is equipped with a time-delay opening interlock, for example of the pneumatic type. When contactor M closes, its normally closed interlock interrupts the coil circuit of contactor DB. This normally closed interlock opens before the timing interlock closes. A semiconductor rectifier is included in the starter to provide a source of d-c excitation. When the stop button is pressed, line contactor M drops out and closes its normally closed interlock. A circuit is then established to energize the coil of contactor DB, which connects the rectifier to the a-c line. The d-c output terminals of the rectifier are connected to two lines to the motor, thereby applying d-c excitation to the motor stator. The motor will be retarded to standstill. After a specified interval, so adjusted as to permit stopping of the motor, the timing interlock on M opens and disconnects contactor DB. Direct-current excitation is removed, and the motor is ready for a new start.

REDUCED-INRUSH STARTING

Any present-day, general-purpose squirrel-cage motor is capable of being started by applying full voltage to its terminals. However,

REDUCED-INRUSH STARTING

depending on the application of a motor, it may be desirable to reduce the starting torque or starting inrush current, or both. A reduction in starting torque may be desirable in order to reduce the stress on material within the machine, as is the case in some textile-machine applications. Reduction in starting current may be desirable in order to reduce the drop in power-line voltage due to starting current. On a weak power system or on a long feeder, the starting of a large motor may reduce the line voltage to the point where operation of other motors on the same circuit may be impaired. Motors and control are guaranteed to operate successfully as long as line voltage does not drop below 90 per cent of its rated value. Caution should be exercised if motor starting inrushes reduce line voltage below that value.

A peculiar situation exists when motors are connected to combination light and power systems as they exist in downtown areas, commercial buildings, and the like. Voltage fluctuations due to motor starting currents may cause variations in the intensity of lights connected to the same circuit, which are called "light flickers." Whereas exclusive power circuits in a large plant are solely under the control of the plant owner, and he makes decisions on tolerable voltage drops, combination light and power circuits are generally owned by public utilities. Since a multiplicity of subscribers are generally connected to the same power and light circuit, the power company has a responsibility to protect its customers against annoyance due to excessive voltage fluctuations. For this reason, power companies establish rules governing the largest motor which they permit to be equipped with full-voltage starting at any location in their power system.

Power company rules take into consideration the degree of annoyance which light flickers may cause. Figure 24-2 shows the results of a study which was conducted by the Edison Electric Institute to establish the borderline for visibility and irritation by lamp flickers. Visibility as well as irritation depend not only on the magnitude of the voltage fluctuation, but also on the frequency with which these fluctuations occur. The permissible voltage fluctuations based on light flickers are considerably smaller than the 10 per cent reduction in line voltage which can be tolerated by the motors. Therefore, power company restrictions on full-voltage motor starting on their combination light and power circuits are considerably more severe than restrictions which may be imposed on exclusive power circuits within a plant.

CONTROL OF SQUIRREL-CAGE MOTORS

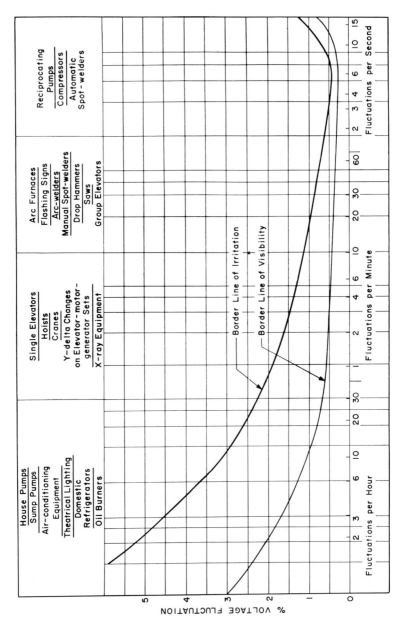

Fig. 24-2 Limits of visibility and irritation of light flickers due to voltage fluctuations.

REDUCED-VOLTAGE STARTING

Starting current and starting torque can be reduced by reducing the voltage applied to the motor stator terminals. A change in voltage applied to the stator terminals results in a change of flux proportional to primary voltage. Likewise, stator and rotor currents vary in proportion to applied stator voltage. Since motor torque is proportional to the product of flux and rotor current, torque is proportional to the square of the voltage applied to the stator. In Fig. 25-2, three commonly used methods of reduced-voltage starting are illustrated, and resultant speed-torque curves are plotted for a typical motor.

An autotransformer is the most expensive but also the most effective means of reducing the starting current. During the whole starting cycle, the terminal voltage at the motor is reduced in proportion to the turn ratio of the transformer. Except for drop in the transformer, the terminal voltage stays constant. Starting torque and maximum

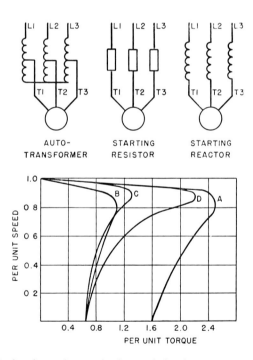

Fig. 25-2 Methods of starting squirrel-cage induction motors. *A*. Full-voltage starting. *B*. Autotransformer, 0.65 voltage tap. *C*. Starting resistor, 0.65 motor standstill voltage. *D*. Starting reactor, 0.65 motor standstill voltage.

torque are reduced proportionally to the square of the turn ratio of the autotransformer. The motor starting current, measured in the motor stator winding, is reduced proportionally to the turn ratio of the autotransformer. Therefore, the starting current, measured in the power line, is reduced proportionally to the square of the autotransformer turn ratio.

Another means of reducing starting current is the primary starting resistor. The resistance drop due to motor current reduces the motor terminal voltage. To determine the desired motor voltage at standstill, it is necessary to know its standstill or locked-rotor current and power factor. The drop through the resistor must be subtracted vectorially from line voltage in order to determine the voltage appearing at the motor terminals. Starting torque is proportional to the square of the ratio of motor standstill voltage and rated motor voltage. Starting current is directly proportional to the ratio between motor-standstill terminal voltage and rated motor voltage. In terms of line current, primary starting resistance obtains less reduction in starting current than the autotransformer for a given reduction in starting torque. While the motor starts and its speed increases, the stator current decreases, and with it the drop through the starting resistor. Therefore the motor terminal voltage rises, and maximum torque of the motor is higher than with the autotransformer starting method.

A starting reactor obtains results similar to a starting resistor. At standstill, the drop through the reactor reduces the motor standstill voltage, the amount of which is determined by subtracting the drop through the reactor vectorially from line voltage. Since the starting power factor of most motors is quite low, fewer ohms are required in a starting reactor than in a starting resistor to obtain a given standstill motor voltage. Starting torque is proportional to the square, and starting current is directly proportional to the ratio between motor terminal voltage and rated motor voltage. As the motor accelerates, its current decreases and its power factor increases. Thus, the drop across the reactor swings out of phase with the drop across the motor. Motor terminal voltage rises above the value obtained with a starting resistor at any given motor speed, and maximum torque of the motor is greatly increased thereby.

To determine the amount of primary resistance or primary reactance to be used for reduced-voltage starting, the locked-rotor impedance of the motor must be known. In Fig. 26-2A, the locked-rotor impedance, resistance, and reactance of the motor are indicated vectorially.

REDUCED-VOLTAGE STARTING

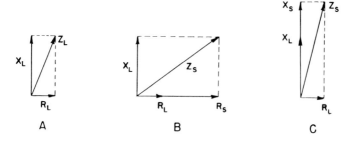

Fig. 26-2 Primary impedance diagrams for reduced-voltage starting. *A.* Locked-rotor motor impedance. *B.* With primary starting resistor. *C.* With primary starting reactor.

It is customary for the motor designer to state the locked-rotor current and the locked-rotor power factor.

The locked-rotor impedance of a three-phase motor is

$$Z_L = \frac{V}{I_L\sqrt{3}} \qquad (11) \bullet$$

where Z_L = locked-rotor impedance
V = rated motor primary voltage
I_L = locked-rotor current

If the locked-rotor power factor is known, locked-rotor resistance R_L and locked-rotor reactance X_L can be determined.

To calculate the amount of primary resistance or reactance to be used, it is necessary to assume the desired starting current. Starting current is proportional to the standstill voltage applied to the motor terminals. With an assumed starting current, the necessary starting impedance is calculated from

$$Z_S = \frac{V}{I_S\sqrt{3}} \qquad (12) \bullet$$

where Z_S = starting impedance
I_S = starting current

The condition existing in primary-resistance starting is illustrated in Fig. 26-2B. In order to obtain a starting impedance Z_S, the starting resistance must be

$$R_S = \sqrt{Z_S^2 - X_L^2} \qquad (13) \bullet$$

CONTROL OF SQUIRREL-CAGE MOTORS

R_S includes locked-rotor resistance R_L. Hence the external resistance required for reduced-voltage starting is $R_S - R_L$.

For primary-reactance starting, a similar condition exists, as illustrated in Fig. 26-2C. To simplify calculations, the resistance of the primary reactor winding is neglected. The error thus introduced would cause the starting current to be slightly reduced below the calculated value; but the actual deviation from the calculated starting current is generally not greater than the deviation which would have to be expected, as a result of tolerances in the design and manufacture of reactors. To realize a desired starting impedance Z_S, a starting reactance

$$X_S = \sqrt{Z_S^2 - R_L^2} \qquad (14) \bullet$$

is needed. The external starting reactor should then have a reactance of $X_S - X_L$.

Primary-resistance and primary-reactance starting are less expensive than autotransformer starting. Primary-resistance and primary-reactance starting is used primarily when the emphasis is on reduction of starting torque. Autotransformer starting is preferred when the emphasis is on reduction of starting current drawn from the power system.

AUTOTRANSFORMER STARTERS

Typical connections of an autotransformer starter are shown in Fig. 27-2. Standard autotransformers are equipped with either two taps for 65 per cent and 80 per cent of line voltage, or with three taps for 50 per cent, 65 per cent, and 80 per cent of line voltage. These taps permit adjustment of the starting current at the installation to suit the particular application conditions.

Autotransformer starters contain two contactors. The run contactor has three poles and is selected as a line contactor for the full motor rating. The start contactor has five poles. Three poles are used to connect the autotransformer to the line; the other two poles are located in the connection between the autotransformer taps and the motor. When the motor is connected to the line on the run connection, it is important that the motor be disconnected from the autotransformer. If this is not done, circulating currents will flow in the autotransformer windings between the transformer taps and the neutral point, causing severe overheating of the transformer windings. In starters up to size 4, the start contactor is the same size as the run contactor; on larger

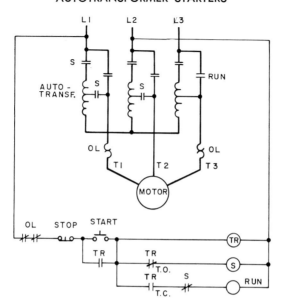

Fig. 27-2 Connections of autotransformer starter with open-circuit transition. S = start contactor; RUN = run contactor; TR = accelerating relay; OL = overload relays.

starters, the start contactor is one size smaller than the run contactor.

Control sequence is as follows: Pressing the start button energizes accelerating relay TR, an instantaneous interlock on which seals around the start button. Relay TR is a timing relay, such as a motor-driven relay or a pneumatic relay equipped with time-opening and time-closing contacts. When the start button is pressed, start contactor S closes instantly, connecting the autotransformer to the line and the motor to taps on the autotransformer. Reduced voltage is now applied to the motor terminals and the motor starts, drawing a current smaller than its locked-rotor current. After a time interval, depending on the setting of relay TR, its normally closed contact opens and its normally open timing contact closes. First, contactor S drops out and, through a normally closed interlock on S, contactor RUN closes, connecting the motor to the line and applying full line voltage to the motor terminals. With this type of starter, it is essential that the start contactor drops out before the run contactor closes; otherwise heavy circulating current transients would occur in the autotransformer, causing high mechanical stresses in the transformer windings.

52 CONTROL OF SQUIRREL-CAGE MOTORS

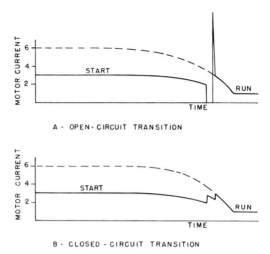

Fig. 28-2 Starting currents of autotransformer starters with open-circuit and closed-circuit transition.

Starting current as a function of time is indicated in Fig. 28-2A. A motor having a locked-rotor current of six times normal motor current is assumed. On the start connection, the current, as seen from the line, is cut approximately in half. When the start contactor drops out, the motor is momentarily disconnected from the line and the current drops to zero. When the run contactor closes and connects the motor to the line, a high transient current peak may occur. A starter which open-circuits the motor winding during the transition from start to run connection is said to have "open-circuit transition." The high transient inrush current peak characteristic of open-circuit transition may be objectionable because of lamp flicker or excessive drop, which reduces motor starting torque to an undesirably low value. This can be avoided by the use of a starter with "closed-circuit transition," which means that the motor remains connected to the power source while transferring from the start to the run connection.

Connections of a typical closed-circuit autotransformer starter are indicated in Fig. 29-2. This circuit is also known as the "Korndorfer connection," after its inventor. This starter contains three contactors. The motor is permanently connected to taps on the autotransformer. Three-pole start and run contactors serve to energize the motor through the autotransformer or directly from the line. In addition, a two-pole contactor is connected in the neutral of the autotrans-

AUTOTRANSFORMER STARTERS

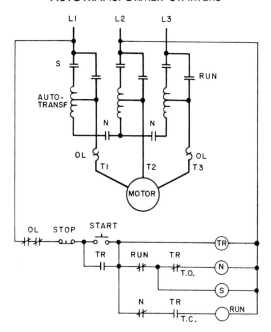

Fig. 29-2 Connections of autotransformer starter with closed-circuit transition (Korndorfer connection). S = start contactor; N = neutral contactor; RUN = run contactor; TR = accelerating relay; OL = overload relays.

former. For starters size 5 and larger, the neutral contactor is also one size smaller than the run contactor.

Control sequence is as follows: When the start pushbutton is pressed, accelerating relay TR is energized and seals itself in through its instantaneously closing interlock. At the same time, contactors S and N close. The motor is connected to the line through taps on the autotransformer, and is thus energized with reduced voltage. After a time interval, governed by the setting of relay TR, a normally closed interlock on TR opens and drops out contactor N. The autotransformer neutral connection is now opened, and its three windings are connected in series with the motor windings, just like a starting reactor. This causes the starting current to rise; but since the motor is not open-circuited, no switching transient occurs. At the same time another interlock on TR closes and, through a normally closed interlock on N, contactor RUN is energized. It connects the motor across the line and short-circuits the autotransformer windings. No circulat-

54 CONTROL OF SQUIRREL-CAGE MOTORS

ing current through the transformer windings can flow, since the neutral connection is opened. A normally closed interlock on RUN drops out contactor S, and the starting sequence is complete.

The effect of the closed-circuit transition on the starting current is illustrated in Fig. 28-2B. When the neutral contactor opens, the motor starting current jumps somewhat, but without switching transient. A second jump in current occurs when the run contactor closes; but again this is not accompanied by any switching transient peak. A comparison of sketches A and B clearly indicates the advantage of closed-circuit transition, although this advantage is partly offset by the higher cost of starters having this type of transition. Because of their lower cost, starters with open-circuit transition are extensively used for small and medium-size motors, whereas those with closed-circuit transition are generally used for larger and more expensive motors.

PRIMARY-RESISTOR STARTERS

Typical connections of a primary-resistor starter are illustrated in Fig. 30-2. Two three-pole contactors are used. An accelerating contactor connects the motor to the line through the starting resistor, and a line contactor connects the motor across the line. The line con-

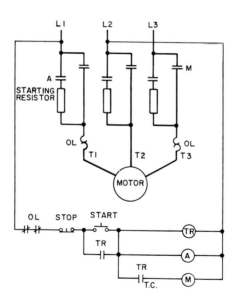

Fig. 30-2 Connections of primary-resistor starter. M = line contactor; A = accelerating contactor; TR = accelerating relay; OL = overload relays.

PRIMARY-RESISTOR STARTERS

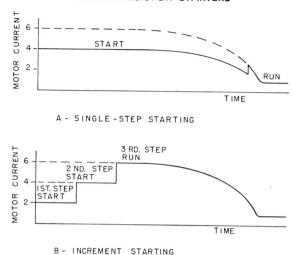

Fig. 31-2 Starting current of single-step and increment primary-resistor starters.

tactor is selected for the full motor rating. On starters up to size 4, the accelerating contactor is of the same size as the line contactor. From size 5 up, the accelerating contactor is one size smaller than the line contactor.

When the start button is pressed, accelerating relay TR is energized and seals itself in. At the same time accelerating contactor A closes and connects the motor to the line with the starting resistor in series. After a time interval, as governed by the setting of relay TR, a time-closing contact on TR energizes line contactor M. The starting resistor is short-circuited, and full line voltage is applied to the motor.

Throughout the whole starting sequence, the motor windings are not open-circuited; therefore, the primary-resistor starter inherently operates with closed-circuit transition. No switching transient current peak may flow. This behavior is illustrated in Fig. 31-2A. The starting current is generally somewhat higher than with an autotransformer starter. When the transfer from the start to the run connection takes place, the jump in current is comparatively small.

When large motors are started on a low-voltage network, the power company sometimes limits the amount of current increase which may be added to the network in any one step. That is to say, the power company does not restrict the total starting current which may be drawn, but it does restrict the "increment" in current which may be imposed on the network on any one starting step. Several such increments can be added to the network consecutively with time intervals between increments. This procedure is established to permit the induction voltage regulators in the network to correct for the increased current resulting from each increment.

Starters containing several steps of primary starting resistance, which are short-circuited by several contactors in time sequence, are called "increment starters" or "network starters." Connections are similar to those shown in Fig. 30-2, except that several accelerating contactors and relays are used. The behavior of an increment starter is illustrated in Fig. 31-2B. The amount of current which may be added on each starting step is limited, in the example shown, to approximately twice normal motor current. Time intervals are provided between steps. During the start, the total starting current of the motor is permitted to rise to its full across-the-line starting current value.

On starters for large motors, reactors are substituted for resistors in order to reduce the losses and consequent heating during starting. Reactors are also preferred for reduced-voltage starting of high-voltage motors because they can be insulated for high voltage more easily than resistors. Connections and control sequence are the same as for primary-resistor starters.

PART-WINDING STARTING

When the stator winding of a squirrel-cage motor is divided into several parallel paths, with the terminals of each section available for external connection, part-winding starting offers an economical means of reducing the starting current with a minimum number of control components. The various sections of the stator winding are connected to the line in sequence, with time intervals between steps, and the starting current is increased in increments until the motor is connected across the line. Small and medium-size motors are available with two parallel sections of stator winding, large motors have been built with three parallel sections.

Motors designed for part-winding starting in two steps have their windings subdivided into two parallel sections with six terminals brought out. Figure 32-2 shows the connections of a combination part-winding starter.

This starter contains two three-pole contactors, each connecting one half of the motor winding to the line. This starting method is known as the "one-half winding" starting method. Pressing the start

PART-WINDING STARTING 57

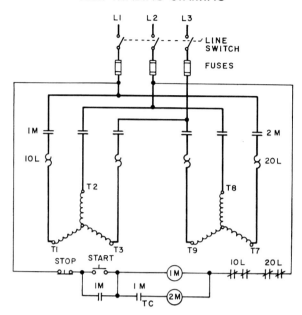

Fig. 32-2 Connections of combination part-winding starter using 3/3 pole contactors for one-half winding starting. $1M$ = first-step line contactor; $2M$ = second-step line contactor; $1OL, 2OL$ = overload relays.

button, contactor $1M$ is energized and connects one half of the motor winding to the line. Contactor $1M$ is equipped with a time-closing interlock (marked TC in Fig. 32-2) which, after a definite time, energizes contactor $2M$, which in turn connects the second half of the motor winding to the line. The motor is now connected across the line on the running step. To provide overload protection during starting and running, overload relays are connected in the two branches of the motor winding. These relays are selected on the basis of one half of the motor horsepower rating. It is important to note that fuses or circuit breakers in combination starters are selected on the basis of full motor horsepower rating. For reasons which will be explained in greater detail in Chapter 7, part-winding starters must be protected by circuit breakers or time-lag fuses having a rating of not greater than twice motor full-load current.

Standard general-purpose motors are built with reconnectable windings for service on either 220 or 440 volts, with connections of primary windings as indicated in Fig. 5-2. Such motors have inherently two

58 CONTROL OF SQUIRREL-CAGE MOTORS

parallel winding sections, they lend themselves naturally to part-winding starting. The need for reduced-inrush starting is particularly pronounced for combination power and light circuits. These are mostly 240/120 volt or 208/120 volt three-phase, four-wire circuits. Part-winding starting has therefore become quite popular for air conditioning and similar applications in commercial buildings which are generally served by combination power and light feeders.

Connections in Fig. 32-2 are indicated for a wye-connected motor. When motors with delta-connected primary winding are built especially for part-winding starting with six leads brought out, external connections are the same as for wye-connected motors. When reconnectable motors for 220/440 volt service have delta-connected windings with nine leads brought out, they may or may not be suitable for part-winding starting. The motor manufacturer's recommendations should be obtained. Special connections of the motor stator winding may be necessary to obtain suitable starting characteristics.

Connecting one half of the motor winding to the line results in a reduction of starting torque and starting current. Motors of different manufacturers vary somewhat in the amount of this reduction on the first starting step. If close guarantees have to be met, the exact figures for a particular application should be obtained from the motor manufacturer. On an average, it can be assumed that starting torque will be approximately 45 per cent of normal starting torque and starting current approximately 65 per cent of normal motor locked-rotor current.

In Fig. 33-2, torque and current characteristics are plotted for a motor having a normal locked-rotor current of six times full-load current. Because of second harmonics, the torque curve has a pronounced dip at one half speed. The magnitude of this dip varies some-

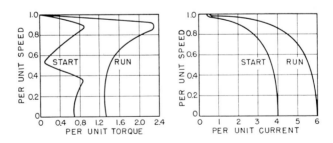

Fig. 33-2 Torque and current characteristics obtained with one-half winding, part-winding starting.

PART-WINDING STARTING

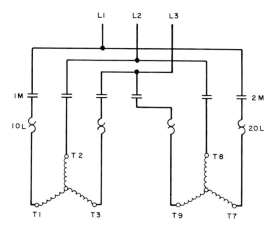

Fig. 34-2 Motor circuit connections of part-winding starter, using 4/2 pole contactors for two-thirds winding starting. $1M$ = first-step line contactor; $2M$ = second-step line contactor; $1OL, 2OL$ = overload relays.

what with motors of different size, but the dip is very pronounced on all motors. Torque may even drop to zero or become negative, which means that, even with a comparatively light load, the motor is likely not to accelerate beyond half speed on the starting step. Considerable noise and vibration must be expected, and the timing interlock for the transfer to the running step should be set for as short a time as possible.

This severe torque dip at half speed during starting can be substantially reduced by energizing two thirds of the motor winding on the starting step. Connections for the motor power circuit of a two-thirds winding part-winding starter are indicated in Fig. 34-2. Control connections are the same as shown in Fig. 32-2. This starter also contains two contactors; however, contactor $1M$ has four poles, whereas contactor $2M$ has two poles. When the start contactor closes on the first step, two winding branches are energized, namely $T1$-$T3$ and $T8$-$T9$. These two branches produce a somewhat unsymmetrical flux pattern, similar to a two-phase winding in Scott connection. On the second step, contactor $2M$ closes and connects the remaining sections of the motor winding to the line, thus producing a symmetrical full-voltage connection.

A part-winding starter is shown in Fig. 35-2. On the left is the four-pole contactor $1M$, at the bottom of which is assembled a pneumatic time-closing interlock. The two-pole contactor $2M$ is to the right. Four overload relays, two for each contactor, are assembled to the sides of the two contactors.

Torque and current characteristics obtained with the two-thirds winding starting method are plotted in Fig. 36-2. The torque characteristic has a considerably less pronounced torque dip at half speed. Thus a motor may be expected to accelerate to nearly full speed. The magnitude of the starting torque is approximately the same as that obtained with the one-half winding starting method, or approximately

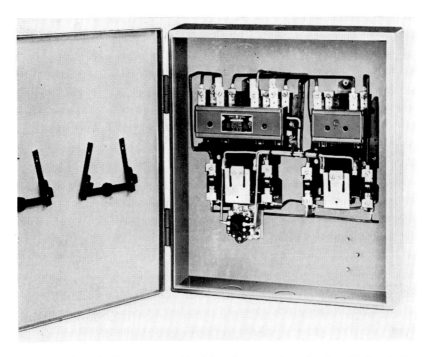

Fig. 35-2 Part-winding starter with 4/2 pole contactors for two-thirds winding starting.

45 per cent of normal starting torque. The current characteristic is the same as that shown in Fig. 33-2 for the half-winding starting method, indicating a reduction in starting current to approximately 65 per cent of normal motor locked-rotor current. This current is the average of the three phase currents. Since the motor is not loaded symmetrically, one phase carries a higher current than the other two phases. The starting current in one phase is reduced to approximately 78 per cent, and the current in the other two phases to approximately

PART-WINDING STARTING

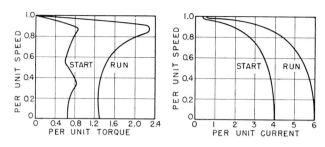

Fig. 36-2 Torque and current characteristics obtained with two-thirds winding, part-winding starting.

55 per cent of locked-rotor current. If the power company considers the current in the high phase as the permissible starting current, rather than the average current, the reduction in starting current obtained with the two-thirds winding starting method may not meet power-company restrictions on starting current.

Part-winding starters with 4/2-pole contactors, as illustrated in Fig. 35-2, can be used as universal part-winding starters obtaining either one-half winding or two-thirds winding starts. Figure 37-2 illustrates the motor circuit connections of a part-winding starter with 4/2 pole contactor combination used to realize the one-half winding

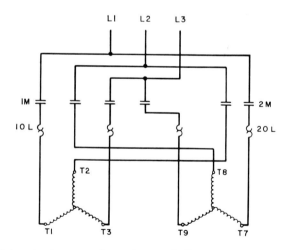

Fig. 37-2 Motor circuit connections of part-winding starter, using 4/2 pole contactors for one-half winding starting. $1M$ = first-step line contactor; $2M$ = second-step line contactor; $1OL$, $2OL$ = overload relays.

62 CONTROL OF SQUIRREL-CAGE MOTORS

starting method. Power connections on the starting step are the same as the ones obtained with Fig. 32-2. A comparison between Figs. 34-2 and 37-2 indicates that internal starter connections are the same, the only difference between the two starting methods being the external connections to the motor.

As compared with reduced-voltage starting methods, part-winding starting has certain advantages. It is less expensive than other methods because it requires no voltage-reducing components such as transformers, resistors, or reactors. It uses only two half-size contactors. The fact that transition from start to run connection is inherently closed-circuit means that there are no high transient current peaks during transfer from start to run connection. On the other hand, a disadvantage of this starting method is the comparatively low starting torque, which cannot be adjusted at the installation since it depends entirely on inherent motor characteristics. Because starting requirements are not severe, part-winding starting has found considerable application in ventilating and air-conditioning equipment for residential or commercial properties served by combination power and light circuits. In view of the wide use of this method, NEMA has established standard ratings for part-winding starters which are listed in Table 3.

The question often arises whether a motor should actually start on the starting step. This is not a requirement when reduced-voltage starting is used. It makes no difference to the power company whether or not the motor turns over and accelerates on the starting step. If the driven machine is capable of being started with a full-voltage starter, it is immaterial whether full motor starting torque is applied initially or after transfer of the control from the starting to the running step.

Table 3. Standard NEMA Ratings for Part-Winding Starters

Size of Starter	Continuous Contactor Rating (amperes)	Horsepower Rating at	
		220 volts	440/550 volts
1PW	27	15	20
2PW	45	30	50
3PW	90	60	100
4PW	135	100	200
5PW	270	200	400

WYE-DELTA STARTING 63

When part-winding starters are applied, it must be considered that, on the starting connection, flux distribution in the stator is not symmetrical and that there will be a certain amount of vibration and noise during starting which tends to reduce the life expectancy of the motor. Part-winding starting is not recommended for drives which are started frequently. The timing of the interlock initiating the transfer to the running connection should be set as short as possible. This is particularly important when the motor fails to accelerate beyond half speed. Increasing the adjustment of the time interlock will not improve the starting performance. Transfer to the symmetrical running condition should be effected as quickly as possible. Because of the unavoidable vibration on the starting step, part-winding starting is not recommended for high-inertia drives requiring a long accelerating time, nor for drives requiring a high starting torque. Part-winding starting is particularly well suited for drives having the characteristic of a fan load.

WYE-DELTA STARTING

Another method of reducing starting current and starting torque by reconnecting the motor windings, without using external starting equipment, is by means of wye-delta starting. This starting method has been used for many years in Europe, but it has only recently found extensive application in the United States. Wye-delta starting requires that the motor be equipped with its stator winding connected in delta, and with six leads brought out. Delta-wound motors have not been common in this country, but in recent years they have been reappraised by motor designers. The interest shown by manufacturers of large centrifugal air-conditioning units in the wye-delta starting method has had a considerable effect on the development of delta-connected motor windings.

Elementary wye-delta starter connections are shown in Fig. 38-2. Two line contactors, $1M$ and $2M$, connect the motor winding to the line in delta connection for running. Each contactor carries 0.577 (1/1.73) motor full-load current. Line contactors of wye-delta starters are suitable for use on motors having 1.73 times the current or horsepower ratings of standard full-voltage starters. A third contactor, S, is used to form the wye point on the starting step. This contactor, which is generally one size smaller than the line contactors, must be capable of handling one third of the rated motor current.

Control sequence is as follows: Pressing the start button energizes

64 CONTROL OF SQUIRREL-CAGE MOTORS

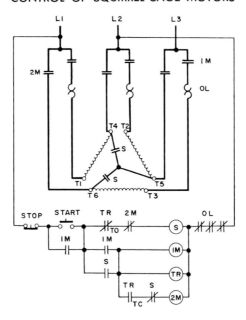

Fig. 38-2 Connections of wye-delta starter with open-circuit transition. $1M$ = first-step line contactor; $2M$ = second-step line contactor; S = wye-point contactor; TR = accelerating relay; OL = overload relays.

contactors $1M$ and S, which connect the motor to the line with the windings connected in wye. Starting current is one third of normal starting current, and starting torque one third of normal motor starting torque. Simultaneously with the contactors, a timing relay TR is energized. After an adjustable time delay, to permit the motor to accelerate in wye connection, a normally closed interlock on TR opens and drops out contactor S, thus opening the wye point of the motor. Another interlock on TR closes, and as soon as S has dropped out, line contactor $2M$ closes. The motor is now connected to the line with its winding connected in delta on the running step.

Three overload relays are shown in Fig. 38-2. These relays are connected in each branch of the motor winding and are selected for 0.577 motor full-load current. This arrangement gives the motor full overload protection, including protection against single-phasing, on the starting and running steps. The use of two overload relays, selected for full-load motor current, in two of the incoming lines would give adequate protection on the running step but not on the starting step. Thus while two relays are permissible under the provisions of the

WYE-DELTA STARTING 65

National Electrical Code, three are preferred because of the higher degree of protection afforded.

A size 6 wye-delta starter is shown in Fig. 39-2. The two bottom bases contain the three-pole line contactors. The two-pole wye-point contactor is on the top base, and a mechanical interlock prevents simultaneous closing of contactors $2M$ and S. The upper base also contains auxiliary relays, overload relays, and a control transformer.

The starter in accordance with Fig. 38-2 includes open-circuit transition between the starting and running steps. Wye-point contactor S opens before line contactor $2M$ closes, thereby interrupting the motor power circuit before the running connection is established. Closed-

Fig. 39-2 Enclosed size 6 wye-delta starter with open-circuit transition. (Courtesy Cutler-Hammer, Inc.)

CONTROL OF SQUIRREL-CAGE MOTORS

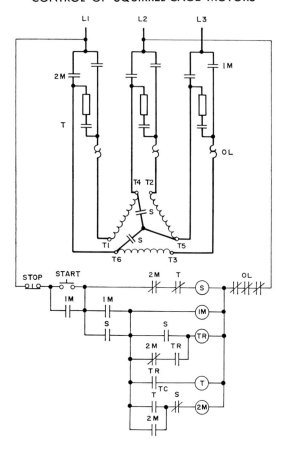

Fig. 40-2 Connections of wye-delta starter with closed-circuit transition. $1M$ = first-step line contactor; $2M$ = second-step line contactor; S = wye-point contactor; T = transition contactor; TR = accelerating relay; OL = overload relays.

circuit transition can be obtained by adding a resistor and a transition contactor, and connections are modified as indicated in Fig. 40-2. A three-pole transition contactor T, added to the conventional wye-delta starter, connects a block of resistance into the motor circuits during transition from starting to running connection.

Pressing the start button energizes contactors $1M$ and S connecting the motor to the line in wye connection on the starting step. Timing relay TR is energized simultaneously. At the end of the timing period, a normally open interlock on TR closes and energizes transition contactor T, which connects a block of resistance between the line and wye-point terminals of the motor winding. After this connection has been established, a normally closed interlock on T drops out the wye-point contactor S. The normally closed interlock on S then closes and sequences-in line contactor $2M$, which establishes the delta connection on the running step. A normally closed interlock on $2M$ deenergizes timing relay TR, which drops out and causes transition contactor T to open. Thus the motor circuit is not interrupted during the transition from the starting to the running step.

The advantage of the wye-delta starting method is the large reduction in starting current, which is one third of the normal motor locked-rotor current. Starting torque is reduced proportionally. This amount is determined by the design of the motor, and cannot be adjusted or modified once the motor is built. The comparatively low starting torque available makes the wye-delta starting method unattractive for drives which have to be started loaded. On the other hand, wye-delta starting is well suited for starting centrifugal compressors used on large air-conditioning equipments which are used in commercial buildings and which are often served by combination power and lighting feeders. The severe reduction in starting current is very attractive for these applications. Experience has shown that such compressors generally accelerate to 95 to 98 per cent of their full-load speed on the starting step, thus the current inrush during transition from the starting to the running step is quite low.

ECONOMIC COMPARISON OF STARTING METHODS

The preceding paragraphs have described the characteristics of various starting methods in engineering terms. While starter selection must be based on technical data, to obtain satisfactory performance, the user is often influenced in his choice of starting method by considerations of cost of starters. Thus although one starting method may have technical advantages over another, these advantages may not be justified in the light of a large differential in starter price.

Figure 41-2 gives a comparison of starter cost to the user for various types of starters and for the more popular starter sizes. The cost of starters providing reduced starting current is expressed as a ratio to the cost of full-voltage starters having the same maximum motor rating. Starter sizes indicated on the abcissa are those of equivalent full-voltage starters. The comparison is based on list prices as given in manufacturers' catalogs. This comparison clearly indicates the high cost differential of reduced-voltage starters in the smaller motor

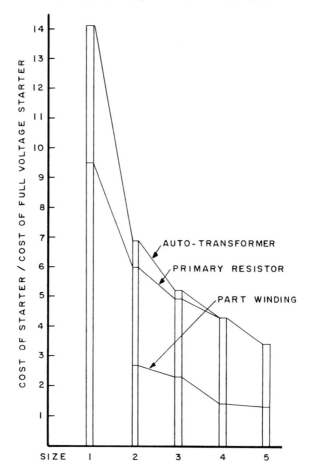

Fig. 41-2 Cost to user of various reduced-inrush starters as compared with full-voltage starters.

ratings, a differential which shrinks as motor ratings increase. The considerable cost advantage of part-winding starters as compared with reduced-voltage starter types is evident.

MULTISPEED MOTORS

Squirrel-cage induction motors are fundamentally constant-speed motors, their speed being determined by line frequency and the number of poles of the stator winding. Specially designed stator windings can be connected for different numbers of poles. Motors equipped

MULTISPEED MOTORS

with such windings are called "multispeed motors" because they can operate at several different speeds. Speed is adjustable in definite steps, the number and spacing of which is determined by the design of the stator winding. At each speed step, the motor will operate as a constant-speed motor.

Two types of stator windings may be used in multispeed motors. Single consequent-pole windings can be reconnected to obtain two different numbers of poles. With this type of winding, the choice of numbers of poles is restricted to a two-to-one ratio. For instance, a consequent-pole winding may be wound for four and eight poles, six and twelve poles, eight and sixteen poles, and so forth. If a different speed ratio is desired, two separate superimposed windings for different numbers of poles may be arranged in the stator slots and connected to the line one at a time. Either one or both of the superimposed windings may be consequent-pole windings. Hence, motors may operate at three or four different speeds. The speeds obtained with the consequent-pole windings must have a two-to-one ratio. Commonly used combinations of stator connections for multispeed three-phase motors are shown in Fig. 42-2.

Multispeed motors are divided into three categories according to the torques developed at the various speeds. Constant-horsepower motors develop the same horsepower at all speeds. Torque on the various speed points varies inversely with the speed. Constant-torque motors develop the same torque on all speed points. Horsepower rating on various speed points varies in proportion to speed. Variable-torque motors develop a torque on each speed point which is proportional to speed. Horsepower rating for various speed points is proportional to the square of the speed.

In Fig. 43-2, speed-torque curves are plotted for typical two-speed motors with a two-to-one ratio for constant-horsepower, constant-torque, and variable-torque design. When the motor is switched from the high-speed to the low-speed winding, the low-speed winding develops a considerable amount of retarding torque which, in constant-horsepower motors, may greatly exceed the maximum torque developed by the high-speed winding. This high retarding torque is accompanied by high stator currents. To avoid dangerous stresses on the driven machinery, it may be necessary to insert buffer resistance in the low-speed winding during the period of retardation from high to low speed. The reduction in retarding torque obtained with a given amount of buffer resistance depends on the design data of the specific

70 CONTROL OF SQUIRREL-CAGE MOTORS

motor. The motor designer should be asked for his recommendations for a given motor application.

When motor connections are switched between speed points, the motor is disconnected from the line and reconnected with a different number of poles. This is inherently an open-circuit transition. Changing the number of poles in the motor causes the flux pattern in the motor air gap to change abruptly. If this change takes place quickly, high transient current peaks due to entrapped flux may be observed. High current transients are particularly severe when switching from a high-

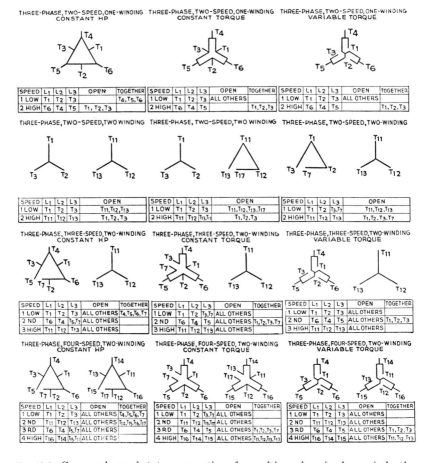

Fig. 42-2 Commonly used stator connections for multispeed squirrel-cage induction motors.

MULTISPEED MOTOR STARTERS

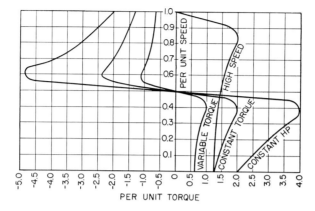

Fig. 43-2 Typical speed-torque curves of two-to-one multispeed squirrel-cage induction motors.

speed to a low-speed connection. In the control operation, as much time as possible should be allowed for the transition between speed points in order to permit decaying of the flux entrapped in the magnetic circuit.

MULTISPEED MOTOR STARTERS

Starters for multispeed squirrel-cage motors are usually full-voltage starters, one contactor being used for connecting the motor across the line on each speed connection. Details of the power circuits depend on the design of the stator windings. Typical power circuit connections for two-speed motors are indicated in Fig. 44-2. When the stator is equipped with two windings, two three-pole contactors are used for connecting each winding to the line, as shown in Fig. 44-2A. The power circuit for each winding is the equivalent of a full-voltage starter for a single-speed motor.

When the motor is equipped with a consequent-pole winding, one three-pole and one five-pole contactor are used. With constant-horsepower design, the slow-speed contactor is the five-pole contactor and the fast-speed contactor is the three-pole one. With constant- or variable-torque design, the slow-speed contactor is the three-pole contactor and the fast-speed contactor is the five-pole one. Connections are indicated in Figs. 44-2B and C. Overload relays are provided to protect the motor at each speed. Stator connections for three-speed motors are combinations of the ones shown in Fig. 44-2,

CONTROL OF SQUIRREL-CAGE MOTORS

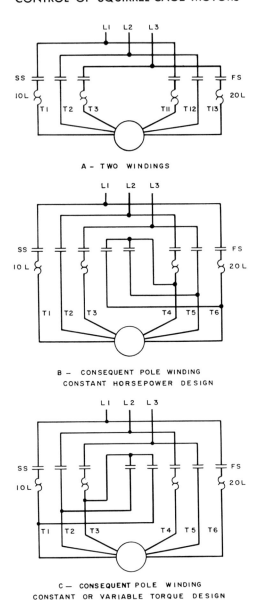

Fig. 44-2 Power-circuit connections of two-speed motor starters. SS = slow-speed contactor; FS = fast-speed contactor; $1OL, 2OL$ = overload relays.

MULTISPEED MOTOR STARTERS

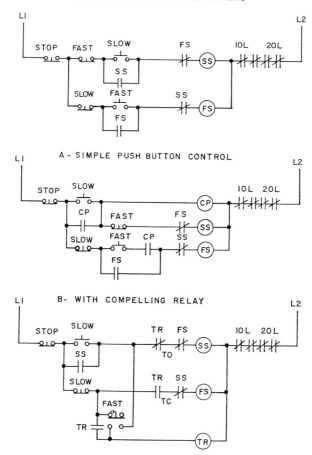

Fig. 45-2 Control-circuit connections of two-speed motor starters. SS = slow-speed contactor; FS = fast-speed contactor; CP = compelling relay; TR = timing relay; $1OL, 2OL$ = overload relays.

with an additional three-pole contactor for the third-speed winding. Connections for four-speed motors are combinations of two sets of circuits as shown in Fig. 44-2.

Various control circuits are in use, their application depending on starting torque and current of the motor under consideration, and the requirements of the drive. Several control circuits are indicated in Fig. 45-2, and any one of the control circuits may be used in connection with any of the power circuits described above.

74 CONTROL OF SQUIRREL-CAGE MOTORS

When starting currents of both stator windings are moderate and the drive may be started on either fast or slow speed, the simple pushbutton circuit of Fig. 45-2A may be used. Pressing the slow or fast pushbutton energizes contactors SS or FS, respectively. When the motor runs at slow speed and the fast pushbutton is pressed, contactor SS drops out and contactor FS closes, and the motor accelerates to its fast speed. Conversely, if the motor runs at fast speed and the slow pushbutton is pressed, contactor FS drops out and contactor SS closes, and the motor decelerates to its slow speed.

For some drives, the starting torque of the fast-speed winding is too low, and the motor must be started on the slow-speed winding before the fast-speed winding can be energized. This sequence can be accomplished by the use of a compelling relay, as shown in Fig. 45-2B. Pressing the slow button closes compelling relay CP, which seals itself in and energizes contactor SS. When the fast button is pressed, contactor SS is dropped out but relay CP stays closed. Contactor FS is energized through an interlock on CP. When the slow pushbutton is then pressed, FS is deenergized and SS closes again. If, however, the fast pushbutton is pressed without the slow button having been pressed, nothing happens, since FS cannot close unless CP has been closed first.

The control circuit just described requires the pressing of two pushbuttons in proper sequence to accelerate the motor to its maximum speed. This inconvenience is avoided by the use of a timing relay, shown in Fig. 45-2C, which obtains progressive acceleration to the high speed. When the slow pushbutton is pressed, contactor SS closes and the motor starts on the slow-speed winding. However, if the fast pushbutton is pressed, timing relay TR is energized and seals itself in through an instantaneous interlock. At the same time, contactor SS is energized through the normally closed timing contact of TR. After relay TR has timed out, its normally closed contact opens, dropping out SS. The normally open contact of TR then closes, causing contactor FS to close. Thus the motor is accelerated in time-controlled progressive sequence to its fast speed by pressing one pushbutton only.

Multispeed motors are reversed by a pair of three-pole reversing contactors, the power contacts of which are connected in series with the pole-changing contactors. Motors may run in either direction at either speed. If the starting current on the fast- or slow-speed windings, or both, is too high, primary resistance and accelerating contactors are added to either or both stator circuits. The additional

SINGLE-PHASING

circuits are the same as described previously for primary resistance starting.

SINGLE-PHASING

Single-phasing of a three-phase motor is not a deliberate control action, but occurs accidentally by interrupting the power flow through one phase to the motor. This loss of one phase may be caused by the blowing of a fuse, or it may be the result of breakage or burning of a power conductor. Single-phasing, in connection with motor protection, is treated in considerable detail in Chapter 6.

To protect a motor against single-phasing, it is necessary to analyze how it affects motor performance. When one phase of the three-phase power supply is interrupted, line voltage appears across two of the stator leads and the motor operates as if it were connected to a single-phase system. A single-phase power system can be considered by symmetrical components as being the resultant of two balanced polyphase systems of opposite phase sequence. Each of the polyphase systems develops torque in the motor, one torque in the positive and one in the negative direction of rotation. The net torque is the difference between the positive and the negative torque at each speed point.

The torque characteristic of a single-phased motor is shown in Fig. 46-2. The maximum torque which can be expected under single-phase operation is between half and two thirds of the breakdown torque obtained during three-phase operation. The breakdown torque of design B motors, which constitute the bulk of industrial motors, is not less than twice normal torque. Therefore, the maximum torque of the motor in a single-phased condition can be expected to be somewhat larger than full-load torque. Hence, if the motor is single-phased while it is running, it will continue to run and carry normal load.

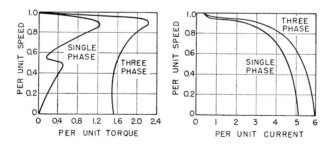

Fig. 46-2 Torque and current characteristics of a single-phased squirrel-cage motor.

CONTROL OF SQUIRREL-CAGE MOTORS

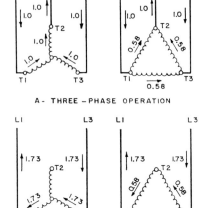

Fig. 47-2 Currents in lines and motor windings during three-phase and single-phase operation of squirrel-cage motor at full load.

A single-phase motor does not develop starting torque. Therefore, if a motor is connected to a single-phased line at standstill, it will not start but will stay on the line stalled. If a motor is unloaded and single-phased while it is accelerating, it will attempt to continue acceleration to full running speed. However, the speed-torque characteristic has a pronounced dip at approximately half speed on account of second harmonics. Therefore, a motor which is single-phased while accelerating may hang at half speed.

Figure 46-2 also indicates the current under single-phase operation, as compared with normal three-phase operation. If the motor is lightly loaded, so that its slip is not increased significantly, the current drawn from the line during single-phase operation is 1.73 times the current drawn during three-phase operation at the same torque. However, when the motor operates close to full load, it runs at higher slip, and the current drawn during single-phase operation is increased somewhat over the theoretical value of 1.73 times the three-phase load current. It may run as high as 2.25 times the current drawn under normal three-phase operation.

When the motor is stalled, the locked-rotor impedance per phase does not differ significantly whether the motor is connected to a single-phase or a three-phase supply. Therefore, the locked-rotor current drawn from the line during single-phase operation bears a ratio of

$1.73/2 = 0.866$ to the locked-rotor current drawn during single-phase operation. If an average locked-rotor current of six times motor full-load current is assumed, the locked-rotor current under single-phase conditions is of the order of 5.2 times rated motor full-load current.

The currents cited in the preceding paragraphs are the currents which flow from the power supply to the motor. The question then arises, what currents are actually flowing in the motor windings during single-phase operation? To illustrate the relation between line currents and currents in the motor windings, Fig. 47-2 indicates the currents flowing in a wye- and a delta-connected motor winding during three-phase and single-phase operation.

In a wye-connected motor, the currents flowing in the windings are the same as the line currents during three-phase as well as single-phase operation. Thus any relays connected in the lines to the motor carry the same current as the motor windings. Hence there is no difference in the response of relays connected in the line with respect to current flowing in the motor windings. With a delta-connected winding, however, a shift in current distribution occurs when the motor is single-phased. During three-phase operation, the three branches of the winding carry the same amount of current, which bears a fixed ratio to the current flowing in the lines to the motors. When the motor is single-phased, one leg of the winding carries twice as much current as the other two legs. Thus the current in the one leg is increased by a ratio of $2/1.73 = 1.16$, as compared with the increase in line current. This means that one leg of the winding carries, comparatively, 16 per cent more current than relays in the lines to the motor would see. In order to avoid possible overloading and consequent overheating of delta-connected motor windings during single-phase operation, it has become customary in the United States to design three-phase industrial motors with delta-connected stator windings having a maximum current density in the conductors which is only 0.866 of the limiting current density used in the design of wye-connected motors. This practice is not necessarily followed by designers of foreign-built motors, so that this design feature has to be considered when applying controllers built in the United States to foreign-built motors.

BIBLIOGRAPHY

1. T. C. Johnson (Editor), *Electric Motors in Industry*, Chapter 4, John Wiley and Sons, New York, 1943.

78 CONTROL OF SQUIRREL-CAGE MOTORS

2. P. B. Harwood, *Control of Electric Motors,* Chapter 16, John Wiley and Sons, New York, 1952.
3. R. W. Jones, *Electric Control Systems,* Chapters 4 and 11, John Wiley and Sons, New York, 1953.
4. W. I. Bendz, "Comparison of Methods of Stopping Squirrel Cage Induction Motors," *Transactions AIEE,* September 1938.
5. O. G. Rutemiller, "Some Basic Circuits for Machine Control," *Electrical Manufacturing,* July 1940.
6. E. H. Hornbarger, "Electric Braking Methods, Reversal of Power by Plugging," *Product Engineering,* July 1944.
7. L. H. Berkley, "Types and Performance of Brakes for A.C. Motors," *Product Engineering,* March 1944.
8. H. Littlejohn, "Basic Control Circuits for Squirrel Cage Motors," *Product Engineering,* June 1946.
9. R. B. Immel, "Interlocks," *Machine Design,* April 1948.
10. H. L. Lindstrom, "Cushion Starting," *Machine Design,* July 1951.
11. G. W. Heumann, "Methods for Starting Squirrel Cage Motors," *Mill and Factory,* March 1952.
12. F. A. List, "Effects of Rotor Design on A-C Motor Performance," *Product Engineering,* July 1952.
13. R. C. May, "Starters for Squirrel-Cage Induction Motors," *Electrical Manufacturing,* December 1952.
14. G. W. Heumann, "Starters for Industrial Type A-C Motors," *Heating and Ventilating,* April 1953.
15. C. F. Evert, "Dynamic Braking of Squirrel Cage Induction Motors," *AIEE Paper* 54-3.
16. J. C. Ponstingl, "Electric Controls," *The Tool Engineer,* July to December, 1954.
17. G. W. Heumann, "Types of A-C Motor Controllers, Part I," *Mill and Factory,* December 1954.
18. H. W. Cory and T. F. Bellinger, "Selecting Reduced Voltage Methods for Motor Starting," *Machine Design,* August 1955.
19. P. L. Alger and L. M. Agacinsky, "A New Method for Part-Winding Starting of Polyphase Motors," *AIEE Paper* 55-846.
20. J. C. Ponstingl, "Electric Motor Braking," *Machine Design,* January 12, 1956.
21. H. W. Cory and T. F. Bellinger, "Starting Methods for Three-Phase Motors," *Product Engineering,* February 1956.
22. I. W. Best, "Applying Squirrel-Cage Induction Motors," *Electrical Manufacturing,* November 1956.
23. T. F. Bellinger, "Controls for Multispeed Motors," *Product Engineering,* February 1957.
24. F. J. Pallischek, "Fundamentals of Across-the-Line Magnetic Starters," *Mill and Factory,* February 1957.
25. P. L. Alger and J. Sheets, "Better Part Winding Starting," *Power Industry,* October 1957.
26. J. Sheets and W. McMichael, "Applying Part Winding Controllers to Squirrel Cage Motors," *Control Engineering,* July 1958.
27. J. C. Elder, "Basic Electrical Controls," *Automation,* July 1958.
28. J. A. Kilcoin, "Starting Three-Phase Motors," *Machine Design,* August 21, 1958.

29. V. J. Picozzi, "Factors Influencing Starting Duty of Large Induction Motors," *AIEE Paper* 59-38.
30. G. W. Heumann, "Part Winding Starting," *Air Conditioning Heating and Ventilating*, May 1959.

PROBLEMS

1. A squirrel-cage induction motor is rated 40 hp, 1760 rpm, 440 volts, 60 cycles, 40 C rise, full-load current 52 amp, code letter G. The motor has a starting torque of 1.68 per unit and a maximum torque of 2.40 per unit. Determine the following:

(a) What is locked-rotor current of motor when operating at rated voltage, assuming that locked-rotor current of motor is actually at midpoint of possible range of values?

(b) When the motor is connected to a system having an actual utilization voltage of 405 volts, what is the maximum torque, the starting torque, and the locked-rotor current?

(c) What maximum load may the motor carry continuously at 440 volts, and what is motor speed with that load?

2. Discuss the following features of a-c reversing control circuits:
(a) Why are contactor coils electrically interlocked?
(b) What is the reason for potential interlocking?
(c) Is potential interlocking used on high-voltage, oil-immersed contactors?

3. Draw up reversing control circuit for a-c contactors, using two forward-reverse-stop pushbuttons, with interlocking through: (a) back contacts of pushbuttons; (b) interlocks on contactors.

4. A squirrel-cage motor is rated 25 hp, 1764 rpm, 440 volts, three-phase, 60 cycles, full-load current 32 amp. The locked-rotor current is 200 amp at a power factor of 0.36. The starting torque is 1.55 per unit, and the maximum torque is 2.63 per unit. The motor is connected to a system delivering 460 volts, through an autotransformer starter. Determine the starting current drawn from the line, starting torque, and maximum torque while the motor is connected to the autotransformer on the (a) 80 per cent tap; (b) 65 per cent tap; (c) 50 per cent tap.

5. An autotransformer starter is designed for use with a 400-hp, 440-volt, three-phase motor:
(a) What size starting and running contactors are used?
(b) What changes would have to be made to adapt the starter for use with a 150-hp, 220-volt motor?

6. The squirrel-cage motor of problem 4 is to be started by a primary resistor starter. Determine:
(a) The amount of resistance needed to obtain a starting torque of 80.8 lb-ft.
(b) The amount of resistance required to reduce the starting current to 125 amp.
(c) The starting torque which the motor will develop with a starting resistor according to (b).

7. A squirrel-cage motor is rated 250 hp, 873 rpm, 2300 volts, three-phase, 60 cycles, full-load current 58 amp. The starting torque is 1.40 per unit. The locked-rotor current is 300 amp at a power factor of 0.32. The motor is connected to a power system delivering 2200 volts, through a primary reactor starter. Determine:

80 CONTROL OF SQUIRREL-CAGE MOTORS

(a) The inductance required to reduce the starting current to 190 amp.

(b) The starting torque which the motor will develop.

8. Consider control circuits for a-c reduced-voltage starters and discuss the following features:

(a) Why are five-pole contactors used as start contactors on autotransformer starters?

(b) On autotransformer starters, is it permissible to have start and run contactors closed simultaneously?

(c) On primary resistance starters, is it necessary to remove the resistor from the power circuit after acceleration? Explain.

(d) On primary reactor starters, is it necessary to underlap line and accelerating contactors? Explain.

9. Assume that the squirrel-cage motor of problem 4 is equipped with a 220/440 volt reconnectable winding, wye-connected. Determine starting torque and starting current if the motor were started by part-winding starting, using (a) the one-half winding method; (b) the two-thirds winding method.

10. Assume that the squirrel-cage motor of problem 4 is equipped with a delta-connected winding and will be started by a wye-delta starter. What will starting torque and starting current be?

11. Discuss the meaning of open-circuit and closed-circuit transition, and explain their significance in regard to line currents. Explain how closed-circuit transition can be obtained with (a) autotransformer starters; (b) primary resistance starters; (c) part-winding starters; (d) wye-delta starters.

12. A pole-changing motor of the consequent-pole design, having a two-to-one ratio, is rated on its high-speed winding 20 hp, 1165 rpm, 220 volts, 60 cycles. Starting torque is 1.35 per unit, and maximum torque is 2.15 per unit. Determine starting torque and maximum torque on the low-speed winding at 208 volts, 60 cycles, if the motor were designed as (a) a constant-horsepower motor; (b) a constant-torque motor; (c) a variable-torque motor.

13. With reference to starting circuit for multispeed motors of the pole-changing type:

(a) How many contactors, and having how many poles, are required for a three-phase, three-speed, two-winding, variable-torque motor?

(b) How many contactors with what number of poles are required for a three-phase, four speed, two-winding, variable-torque motor?

(c) Draw elementary power circuit for the three-phase, four-speed, two-winding, constant-horsepower motor.

(d) Explain the function of a compelling relay.

(e) Explain the function of a progressive-start timing relay.

14. The power supply to the squirrel-cage motor of problem 4 is single-phased:

(a) Assuming that the motor is fully loaded, what line current can be expected? Give a possible range of not less and not more than ....

(b) What maximum torque can the motor be expected to develop?

(c) Assuming that the motor winding is wye-connected, what currents will flow in the three branches of the winding?

(d) Assuming that the motor winding is delta-connected, what currents will flow in the three branches of the winding?

(e) Assuming that the motor is stopped when the line is single-phased, what will the starting torque and current be when the motor is energized?

3 CONTROL OF WOUND-ROTOR MOTORS

In external appearance, wound-rotor motors do not differ greatly from squirrel-cage motors. A cutaway view of a wound-rotor motor is given in Fig. 1-3. The stator is built like that of a squirrel-cage motor, and it carries a winding which is imbedded in the stator slots. However, the rotor also carries a wound winding which is imbedded in

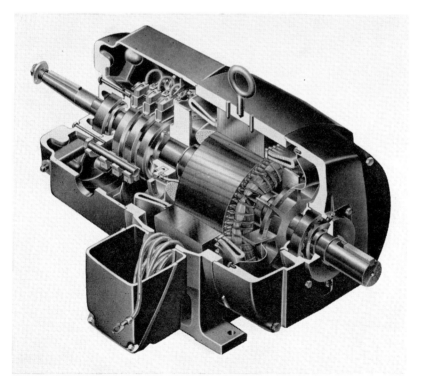

Fig. 1-3 General-purpose wound-rotor induction motor.

82 CONTROL OF WOUND-ROTOR MOTORS

the rotor slots. The three phases of this winding are connected on one side to a common wye point, whereas the other side of each phase winding is connected to slip rings. By means of external switching arrangements, the slip rings can be short-circuited; or external resistance can be connected in series with the rotor winding for the purpose of controlling speed and torque of the motor. Since the stator is the primary circuit of the motor connected to the line, the rotor circuit of the motor is often referred to as the "secondary circuit" of the motor.

WOUND-ROTOR MOTOR CHARACTERISTICS

With its slip rings short-circuited, the speed-torque curve of a wound-rotor motor is very similar to that of a standard normal starting torque squirrel-cage motor. Since the secondary resistance of wound-rotor motors is generally lower than that of squirrel-cage motors, starting torque is lower and starting current is higher.

When resistance is inserted in series with the secondary winding, that is when resistance is connected to the slip rings, the secondary drop is increased and the slip (or rotor frequency) must increase correspondingly, so that the voltage generated in the rotor balances the drop across the internal and external resistance. To calculate speed-torque curves, it is necessary to determine generated secondary voltage, which is a measure of speed, and secondary current, which is a measure of torque. Expressed in per unit,

$$s = E_2 \qquad (15) \blacksquare$$

$$T = I_2 \qquad (16) \blacksquare$$

where E_2 = per unit generated secondary voltage
s = per unit slip
I_2 = per unit secondary current
T = per unit torque

A wound-rotor motor connected to the line with its secondary winding open-circuited acts as a transformer. Secondary voltage is equal to line voltage multiplied by the transformer ratio between primary and secondary windings. No current flows in the secondary, no torque is developed, and the motor remains at standstill. That means, slip is unity.

To express motor operating conditions on a per unit basis, the following base quantities are defined:

WOUND-ROTOR MOTOR CHARACTERISTICS

Base voltage $E_{2\,\text{normal}}$ = standstill voltage between slip rings at full load

Base current $I_{2\,\text{normal}}$ = secondary full-load current per phase

Base torque = rated torque

Base speed = synchronous speed

Total resistance of the secondary circuit consists of the internal resistance of the rotor winding per phase and the external resistance per phase which is connected to the slip rings.

$$R_2 = R_i + R_s \qquad (17)$$

where R_2 = total secondary resistance per phase
R_i = internal rotor resistance per phase
R_s = external secondary resistance per phase

To convert resistance in ohms to per unit value, base resistance is defined as follows:

$$\text{Base resistance } R_{2\,\text{normal}} = \frac{E_{2\,\text{normal}}}{\sqrt{3}\, I_{2\,\text{normal}}} \qquad (18) \bullet$$

With base resistance R_2 connected in the secondary circuit and base current I_2 flowing, the motor develops base torque, secondary voltage equals base E_2, and the slip is unity. In other words, starting torque equals base torque.

Base resistance in ohms is generally obtained from the motor designer. It does not appear on motor nameplates and is not included in motor handbook information. Some motor designers give so-called "rheostat ohms"—that is, external ohms which have to be connected to the slip rings to obtain normal torque at standstill. Rheostat ohms divided by $1 - s$ equals base resistance.

If base resistance or rheostat ohms cannot be obtained from the motor designer, base resistance of a three-phase motor can be calculated from the following formula:

$$R_{2\,\text{normal}} = 249 \times \frac{\text{rated motor hp}}{I_{2\,\text{normal}}^2 (1-s)} \qquad (19) \bullet$$

This formula is based on the following consideration: With the motor standing still and developing normal torque, the ohmic loss in each secondary resistor phase equals one third the rated horsepower output of the motor. Since rated horsepower and secondary full-load current appear on motor nameplates, formula 19 can always be used to determine base resistance.

It is customary to stamp on motor nameplates the open-circuit

CONTROL OF WOUND-ROTOR MOTORS

secondary voltage between slip rings. This is the voltage which would be measured between the slip rings if they were open-circuited, and with rated voltage applied to the motor stator. This voltage is given so that the motor user may know for what voltage he has to insulate the secondary circuit. The open-circuit secondary voltage is somewhat higher than base voltage E_2 normal, since it does not include the drop in the motor stator due to full-load current. If the secondary open-circuit voltage is known, it can be assumed, as a fair approximation, to be 4 per cent higher than base secondary voltage. Hence, secondary open-circuit voltage can be used to determine base resistance.

EFFECT OF SECONDARY RESISTANCE ON MOTOR SPEED

During normal operation, the torque delivered by the motor is well below maximum torque and the speed is above the speed at which maximum torque occurs. Within this range, the speed-torque curves are straight lines, and the relation between speed and torque can be expressed in per unit:

$$s = 1 - S$$
$$= T(R_i + R_s) \qquad (20) \blacksquare$$

For any given amount of torque, slip increases and speed decreases with increasing secondary resistance. If speed versus torque is plotted, the intersection between the speed line and the base torque line is a direct measure of secondary resistance, as indicated in Fig. 2-3.

Starting torque T_s is the torque developed by the motor at standstill or unity slip. When the secondary resistance is high, say 0.5 or more, the reactance of the motor is only a small fraction of the resistance and can be neglected. The speed-torque curve is a straight line between full speed and zero speed. Starting torque is then inversely proportional to secondary resistance. In per unit,

$$T_s = \frac{1}{R_i + R_s} \qquad (21) \blacksquare$$

To lay out a secondary resistor and to plot speed-torque curves of a specific motor, the internal resistance R_i must be known. Expressed in per unit, internal resistance equals the normal slip of the motor at full load. This relationship becomes apparent by substituting $T = 1$ and $R_s = 0$ in formula 20. In Fig. 3-3 internal resistance of standard continuously rated general-purpose motors in per unit is plotted versus motor rating. A band is given between upper and lower limits. Slip

EFFECT OF SECONDARY RESISTANCE ON MOTOR SPEED

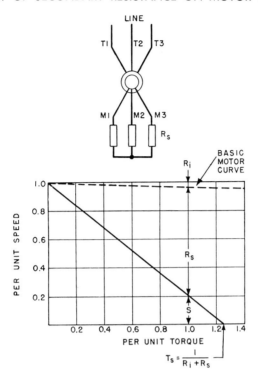

Fig. 2-3 Effect of secondary resistance on speed of wound-rotor induction motor.

increases with the number of poles; it also varies somewhat with motors of different designs and of different manufacture. The lower limit is an average for four-pole motors, and the upper limit is an average for ten- to sixteen-pole motors. For intermittent-rated motors, such as those generally used for material-handling equipment, internal resistance is somewhat higher.

The amount of resistance connected in the secondary circuit has no effect on the maximum torque of the motor. It does, however, affect the speed at which maximum torque occurs. Maximum torque occurs at a speed at which the transfer of power from stator to rotor is maximum. At this point

$$\frac{R_2}{s} = X_1 + X_2 = \text{const} \qquad (22)$$

where X_1 = stator reactance
X_2 = rotor reactance

86 CONTROL OF WOUND-ROTOR MOTORS

Hence the slip at which maximum torque occurs is proportional to the total resistance in the secondary circuit.

In Fig. 4-3, speed-torque curves for a typical wound-rotor motor are plotted for various amounts of total secondary resistance. These curves are given for motoring performance in quadrant 1 and for plugging or countertorque performance in quadrant 4. The amount of external resistance has little effect on speed at light loads. It is impracticable to use wound-rotor motors as adjustable-speed motors for reductions in speed to less than one half at full load because the slope of the speed-torque curves at greater speed reductions becomes too steep, and slight variations in torque cause undesirably high speed changes.

If a greater amount of speed reduction at full load is desired, the motor may be built with a pole-changing stator winding, similar to the windings used on multispeed squirrel-cage motors. It is then practicable to reduce the speed to one half of each no-load speed by inserting secondary resistance. For instance, a motor with a one-to-two speed range by pole changing can be used as an adjustable-speed motor down to one quarter of its high speed. When motor characteristics are reduced to per unit values, pole changing influences the characteristics of the stator winding but not of the rotor winding of a given motor. Pole changing does not affect the per unit secondary data, and a given amount of per unit secondary resistance will obtain the same per unit speed reduction with the high-speed and the low-speed stator connections.

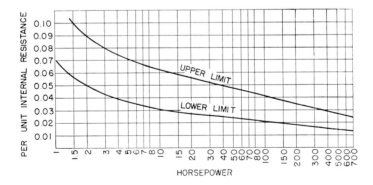

Fig. 3-3 Internal resistance of continuously rated general-purpose wound-rotor induction motors.

UNBALANCED SECONDARY RESISTANCE

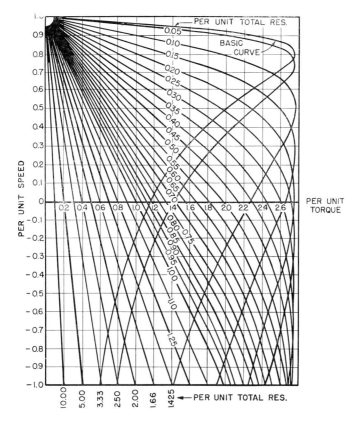

Fig. 4-3 Speed-torque curves of a typical wound-rotor motor with various amounts of per unit secondary resistance.

UNBALANCED SECONDARY RESISTANCE

Thus far, in considering the performance of wound-rotor motors with external secondary resistance, the assumption has been that the same amount of resistance is inserted in each phase of the rotor. In other words, it has been assumed that the rotor circuit is balanced. Using unequal amounts of resistance in the rotor phases has advantages under some operating conditions. Figure 5-3 illustrates commonly used unbalanced rotor connections.

If resistance is connected only in two rotor phases, and the third phase is left open, the effective resistance is twice the actual resistance per phase. This connection is used frequently on the first point of

CONTROL OF WOUND-ROTOR MOTORS

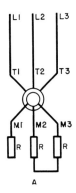
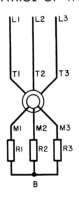

Fig. 5-3 Effect of unbalanced resistance in secondary circuit. A. Leaving one rotor phase open. $R_{\text{eff}} = 2R$. B. Using unequal resistance in the three rotor phases.

manual starters, since it simplifies starter design and cuts in half the amount of resistance required to obtain a given starting torque. Closing the third rotor phase on the second point of the starter balances the rotor circuit and doubles the starting torque obtained on the first point. The unbalance in the rotor circuit causes torque pulsations and additional mechanical stresses in the motor. For this reason starting with one rotor phase open should be used only for motors having a rating of 75 horsepower or less.

When manual drum switches are used to change the amount of resistance in the secondary circuit, design of the switch development can be simplified if resistance is changed in unbalanced steps. The motor speed-torque curve for unequal amounts of resistance in the rotor phases can be calculated with an equivalent effective resistance as follows. Let $R1$, $R2$, $R3$ be the total resistance in each of the three rotor phases. Then

$$R_{\text{eff}} = \frac{R1 + R2 + R3}{3} \qquad (23)$$

Because of the unbalance in the secondary circuit, unbalanced currents of slip frequency flow in the rotor phases and balanced components of currents at two frequencies flow in the stator phases. One is the fundamental line frequency, and the other is a subharmonic frequency which depends on speed. These unbalances cause irregularities in the speed-torque curve of the motor. The starting torque is somewhat lower than would be calculated with an effective resistance of the average resistance according to formula 23. Tests on actual motors indicate that better results are obtained if the starting torque is calculated with an effective resistance

ACCELERATION

$$R_{\text{eff}} = \frac{R1R2 + R2R3 + R3R1}{R1 + R2 + R3} \quad (24)$$

Also, the torque curve has a pronounced dip in the vicinity of half speed. The extent of this dip depends on motor design, the amount of secondary resistance, and the degree of unbalance between the rotor phases. With a large amount of resistance in the secondary circuit, as would be used on the starting point of a controller, the torque dip is not very pronounced and would permit a lightly loaded motor to accelerate to full speed. However, on running points of the controller, the motor may be unable to accelerate past half speed. The unbalance in the rotor phase currents may cause rather severe torque pulsations. It is generally considered good practice for motors larger than 75 horsepower not to be operated continuously with unbalanced rotor connections.

ACCELERATION

To accelerate the motor from standstill to full speed, the starting resistance which had been chosen to obtain a desired starting torque has to be ultimately short-circuited, so that the secondary motor circuit is short-circuited at the slip rings. During acceleration, current and torque peaks must be limited. Limitation of current peaks is necessary to prevent overstressing of the devices used for switching the secondary current, and to limit the voltage drop due to current peaks in the primary circuit. Since motor torque is proportional to the square of the voltage applied to the motor stator terminals, it is important that the motor not be prevented from developing its maximum torque by excessive primary voltage drop. Limitation of torque peaks is important to prevent overstressing the mechanical parts of the driven machinery. To limit current and torque peaks to a reasonable value, it may be necessary to short-circuit the starting resistor in several steps.

For an explanation of wound-rotor motor acceleration, refer to Fig. 6-3. Sketch A is the elementary diagram of a wound-rotor motor circuit with three sections of secondary resistance connected in each of the secondary phases. A starter employing such a resistor arrangement is called a "four-step or four-point starter." On the first step, the motor primary is connected to the lines with all secondary resistance in the circuit. On steps 2, 3, and 4, resistance sections are short-circuited in the secondary circuit by moving the external wye point along taps on the resistor until, on the last step, the slip rings are short-circuited.

CONTROL OF WOUND-ROTOR MOTORS

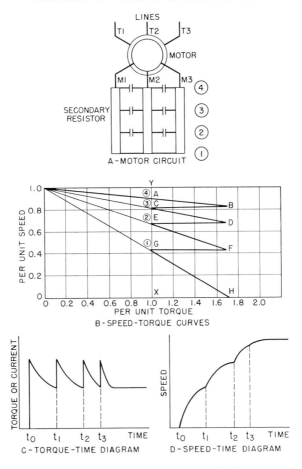

Fig. 6-3 Acceleration of a wound-rotor induction motor.

The switching devices for short-circuiting resistance sections may be the contacts of a manual secondary starting switch, or they may be accelerating contactors. When a manual starting switch is employed, the number of starting steps is the same as the number of mechanical positions through which the switch may be turned. For magnetic starters, the number of steps or points is one more than the number of accelerating contactors used.

In sketch *B*, the speed-torque curves are plotted for a starter in accordance with sketch *A*. The speed-torque curve for each step is a straight line, and the point at which it intersects the unity torque line is determined by the amount of resistance in the motor secondary circuit on that step. The total amount of starting resistance per phase is governed by the desired starting torque, which must be selected by the application engineer to suit the driven machinery. The resistor application data in Chapter 5 of Part I give recommended resistor classifications, which in turn indicate starting torques required for commonly used types of machinery. The heavy lines in sketch B indicate how the motor accelerates on each step, assuming that the load torque of the drive is constant over the whole range from zero to full speed. Furthermore, the resistor layout is such that the torque peaks during acceleration are equal to the starting torque.

When connected to the line, with all resistance in the secondary circuit, the motor develops a starting torque equal to OH. In the per unit system, OX is the full-load torque and XY is base resistance, as defined in formula 18. In per unit, YA is the internal resistance of the motor secondary windings per phase, AG the total external starting resistance per phase, and YG the total secondary resistance per phase. Expressed in per unit, the starting torque is

$$T_{\text{starting}} = OH = \frac{1}{YG} \qquad (25) \blacksquare$$

As the starting torque exceeds the load torque, the motor gains in speed. During acceleration, the torque delivered by the motor decreases. The motor accelerates along speed-torque curve HG until the motor torque becomes equal to the load torque. At this point, motor speed in per unit is XG. Since motor secondary current is proportional to motor torque, the decrease in motor torque corresponds to a decrease in secondary current, which in turn results in a decrease in drop across the starting resistor. Correspondingly, the secondary voltage of the motor and its slip decrease; hence motor speed increases. Expressed in per unit, OH is the starting current and OX the full-load current. Likewise, XY is secondary standstill voltage for unity slip, and YG is secondary voltage and slip corresponding to speed at point G.

Sketch C indicates in a general manner how torque and secondary current vary as a function of time. At time t_0, the motor is connected to the line and starting torque is developed instantly. At time t_1, the torque has decreased to the full-load value. The exact shape of the torque-time curve varies with the mechanical characteristics of the drive, such as friction and inertia, and the sketch intends only to illustrate the general shape of the curve. In a corresponding manner,

92 CONTROL OF WOUND-ROTOR MOTORS

sketch D shows how the motor speed increases during the time interval t_0–t_1.

When equilibrium between motor torque and load torque has practically been reached at time t_1, a section of starting resistance is short-circuited on the second step of the starter. Expressed in per unit, this resistance step is represented by EG in sketch B. Since motor secondary voltage and speed cannot change instantly, secondary current and motor torque must rise to such a value that the drop through the remaining sections of external and internal secondary resistance equals YG. Thus secondary current and torque increase suddenly to point S, along line GF. As the torque then exceeds the load torque, the motor accelerates along line SE. The torque decreases and the speed increases, until at time t_2 the motor torque practically equals the load torque at point E.

The next resistance section is then short-circuited. The torque rises instantly to point D. The motor accelerates along line DC until at point C motor torque equals load torque at time t_3. When the last section of resistance is short-circuited, the motor slip rings are shorted and the only resistance remaining in the secondary circuit is the internal resistance of the rotor winding. The torque rises to point B and the motor accelerates to point A, which is the natural full-load speed of the motor. Acceleration has then been completed.

The required total starting resistance having been determined, it is convenient to determine the resistance to be used on each step by means of a graph similar to sketch B. Since commercial resistors may deviate ±10 per cent from their nominal resistance, there is no merit in calculating a starting resistor layout by a time-consuming analytical method which yields very accurate results on paper but cannot be realized in practice because of inherent variations in resistance of practical resistors.

For a resistor layout, proceed as follows: Determine the desired starting torque from application data, so that point H is established and line 1 can be drawn. The intersection with the unity torque line yields point G. Draw a horizontal line through point G, and make GF equal to XH. Then draw line 2, which fixes point E. Draw a horizontal line through point E, and make ED equal to XH. Line 3 can then be drawn and point C established. Draw a horizontal line through C which intersects the fundamental speed-torque curve in point B. When CB is equal to or smaller than XH, the accelerating peaks do not exceed the starting torque. If it is found that CB is larger than XH, either the torque peaks must be somewhat increased or an additional step of starting resistance must be used. After the speed-torque diagram has been established, the intersections of the speed-torque lines with the unity-torque line give the per unit values of the starting-resistance sections. EG is the first section, CE the second, and AC the third. If many resistor layouts are to be prepared, considerable time can be saved by first preparing a general chart of speed-torque curves similar to Fig. 4-3. By obtaining blueprints of this chart and penciling in accelerating curves corresponding to Fig. 6-3B, it is relatively easy by trial and error to establish a workable layout which will help determine the minimum number of accelerating steps necessary to keep torque peaks at a tolerable value.

MANUAL CONTROL OF ACCELERATION

The principal advantage of the wound-rotor motor is the ability to adjust speed and torque on a multiplicity of controller points and to reduce starting-current peaks considerably more than is practicable with squirrel-cage motors. However, as a result of improvements in motor torque and current characteristics, the squirrel-cage motor has largely replaced the wound-rotor motor as a general-purpose a-c drive. Wound-rotor motors are used today principally for special drives on which speed control is essential. Examples are fans, pumps, and large air-conditioning units. Wound-rotor motors are also applied to large drives, sometimes because it is desirable to control the starting torque, sometimes to reduce starting current peaks to a value which can be tolerated by the power system. Examples are crushers and ball mills. Important applications of wound-rotor motors are material-handling drives such as cranes and hoists. On such drives it is necessary to have control of speed and torque to adjust motor performance to varying loads of material to be handled. While magnetic control has largely superseded manual controllers, because of its greater ease of manipulation by the operator, manual or semimagnetic controllers are still used to some extent on less frequently operated drives because of their lower cost.

The oldest and simplest type of wound-rotor motor controllers are face-plate secondary controllers, as illustrated in Fig. 7-3. The resistor consists of three legs, one for each secondary phase. The spider-like contact arm, also consisting of three legs, forms the wye point. Turning the contact arm moves the wye point along the resistor sections and varies the amount of resistance in each secondary phase. Depending on whether the secondary resistor is designed for short-time or continuous duty, the controller can be used for starting or for

94 CONTROL OF WOUND-ROTOR MOTORS

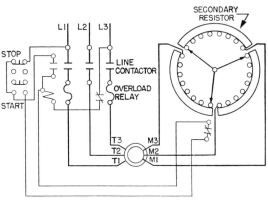

Fig. 7-3 Face-plate secondary controller for wound-rotor induction motor.

speed control. The time allowed for the motor to accelerate and the resulting line current peaks depend entirely upon the skill with which the operator manipulates the controller.

Since the face-plate controller takes care of the rotor connections only, a separate magnetic starter is used for the motor stator connections. The primary magnetic starter shown in Fig. 7-3 is the same as a standard across-the-line starter for a squirrel-cage motor and is selected on the same basis. The primary control is actuated by a start-stop pushbutton station, and overload protection is provided

MANUAL CONTROL OF ACCELERATION

by the overload relays on the starter. When the motor is started, the secondary rheostat must be in its starting position, with all resistance in each rotor phase. An auxiliary contact is closed in the starting position by the contact arm. This auxiliary contact is connected in series with the start pushbutton, so that the line contactor can be closed only when the secondary rheostat is in the starting position.

Although face-plate controllers have the advantage of mechanical simplicity and resultant low cost, they lack sturdiness and are unsatisfactory for frequent operation. This disadvantage is largely overcome by drum switches. The linear sliding motion between contact buttons and contact arm is superseded by the rotary sliding motion between stationary contact fingers and circular contact segments, arranged on a drum.

A typical drum switch for wound-rotor motor control is shown in Fig. 8-3. The segments at the top of the switch serve to connect the motor primary to the line. The lower segments short-circuit sections of the secondary resistor in successive steps. Such switches, reversing and nonreversing, can be obtained for starting duty only, for intermittent service, or for continuous speed control. The drum layout may be such that all positions obtain balanced secondary connections, or that only every other position is balanced.

Compared with face-plate controllers, drum switches have the following advantages:

The handle is easier to manipulate.

Greater flexibility exists in the design of drum developments.

The design of the switch is more compact and results in sturdier mechanical construction.

Contact springs back of the contact fingers maintain higher and more uniform contact force.

Drum switches can be built either for reversing or nonreversing duty.

Drum switches are often used in conjunction with magnetic contactors for primary motor control, thereby facilitating the inclusion of overload protection and the introduction of undervoltage protection and limit switches. Typical applications on which semimagnetic controllers are often used are infrequently operated shop cranes, small mine or construction hoists, and small movable bridges.

The sliding motion between the stationary contact fingers and the drum cylinder causes high contact wear and considerable burning of the contacts when the circuit is opened. This disadvantage is overcome by cam switches. A cam switch resembles a drum switch in overall appearance. However, each circuit is made up of an electrically isolated pair of stationary and movable contacts. Each movable contact is actuated by a set of molded cams which force the contact open and closed. The closing sequence of any contact with respect to the handle position is determined by the shape of operating cams. The advantage of these switches is that sliding contacts are replaced by butt contacts which include a slight sliding motion for self-cleaning, similar to that used in magnetic contactors. Cam switches are built with contact ratings up to 1000 amperes. They are used primarily for the secondary control of wound-rotor motors, principally for large fans and air-conditioning units. The forces necessary to move such

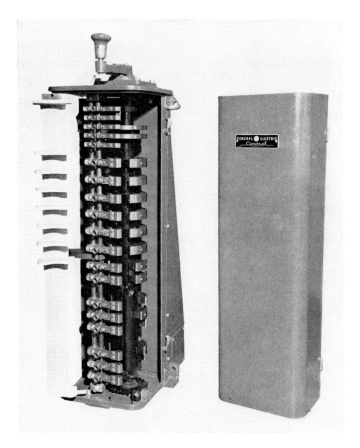

Fig. 8-3 Drum switch with horizontal operating mechanism, cover removed.

MAGNETIC CONTROL OF ACCELERATION

Fig. 9-3 Motor-operated cam switch for wound-rotor motor speed control, rated 1000 amperes, with thirteen balanced positions.

large switches are considerable, and pilot-motor operating mechanisms as shown in Fig. 9-3 are often used on large switches. The pilot mechanism is designed so that the switch can travel only between definite positions, and contacts cannot be partly closed by stopping the switch between positions. Since only small control-circuit devices are required to control the pilot motor, such motor-operated switches lend themselves readily to use in conjunction with magnetic controllers.

MAGNETIC CONTROL OF ACCELERATION

When magnetic control is used, the closing of the secondary accelerating contactors can be accomplished automatically in proper sequence after the line contactor has closed. To obtain smooth acceleration, the accelerating contactors should close when the motor has accelerated on the preceding steps to such a speed that the contactor closing will not cause the torque or the current to exceed a pre-

CONTROL OF WOUND-ROTOR MOTORS

determined peak. Thus in Fig. 6-3 the accelerating contactors should close at or a little after the time intervals $t1$, $t2$, $t3$. Depending on the criterion used for controlling the closing of the accelerating contactors, the following methods of accelerating control are available:

1. *Current-limit control.* Contactor closing is governed by current-limit relays which permit the accelerating contactor of each step to close when the motor current has dropped from a peak value to a preset lower value.

2. *Secondary-frequency control.* As the motor accelerates from standstill to full speed, the frequency of the rotor circuit drops from line frequency to a few cycles. Frequency-responsive relays permit accelerating contactors to close as the motor reaches predetermined speeds.

3. *Definite-time control.* The closing of accelerating contactors occurs in a definite time sequence, the timing being so adjusted as to obtain uniform accelerating current peaks.

In the following paragraphs, the basic connections of starters using the foregoing accelerating control methods are described. The elementary diagrams shown to illustrate the accelerating control include only such circuit details as are necessary to explain the basic control systems. Additional control elements, especially protective devices, are omitted here for the sake of simplicity.

CURRENT-LIMIT ACCELERATION

Current-limit acceleration (also called "series lockout" by some authors) uses relays which measure the stator or rotor current to control the closing of the accelerating contactors. One form of current-limit accelerating controller is shown in Fig. 10-3. Accelerating relays $1AX$ and $2AX$ have two coils, one potential coil and one current coil which is connected to one motor stator phase, either directly or through a current transformer. With no current flowing in the potential coil, the relay contact is open irrespective of current flowing in the current coil. When the potential coil of the relay is energized, the current flowing in the current coil determines whether or not the relay closes its contact. The current coil restrains the potential coil and prevents the relay from closing its contact if the current is high, as during acceleration. The current coil is so calibrated that it permits closing of the relay contact when the motor current has dropped to a predetermined value, usually between 100 to 125 per cent of full-load

CURRENT-LIMIT ACCELERATION

current. Thus, when the potential coil is energized, the relay contact is under the control of the current coil.

With the motor deenergized, potential and current coils of relays $1AX$ and $2AX$ are deenergized, and the contacts of both relays are open. Pressing the start button causes contactor M to close. The inrush current of the motor restrains both relays $1AX$ and $2AX$. Through an interlock on M, the potential coil of relay $1AX$ is connected in series with the coil of contactor $1A$. Since the impedance of the $1A$

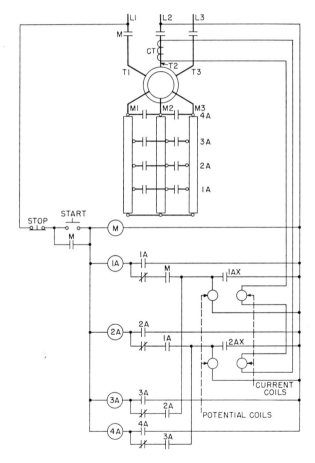

Fig. 10-3 Starting control circuit with current-limit acceleration. M = line contactor; $1A, 2A, 3A, 4A$ = accelerating contactors; $1AX, 2AX$ = accelerating relays; CT = current transformer.

coil is low as compared with that of $1AX$ potential coil, the resultant current is insufficient to close $1A$ but sufficient to pick up $1AX$, provided it is not restrained by the current coil. When the motor current drops to the preset value, the current coil releases the relay contact. Relay $1AX$ closes its contact and connects the $1A$ coil directly across the control source so that $1A$ closes. As $1A$ seals itself in, a normally closed interlock breaks the connection between $1A$ coil and $1AX$ potential coil, so that $1AX$ drops out and opens its contact.

When $1A$ closes, an interlock connects the coil of contactor $2A$ in series with the potential coil of relay $2AX$. The closing of $1A$, however, has caused an increase in motor current, so that the current coil $2AX$ keeps the contact of that relay open. When the current drops again to the preset value, $2AX$ closes its contact and $2A$ closes, sealing itself in and breaking the connection between $2A$ coil and $2AX$ potential coil. At the same time the coil of contactor $3A$ is connected in series with the potential coil of relay $1AX$, but the contact of that relay stays open because of the current peak due to $2A$ closing. As the current drops, $1AX$ permits $3A$ to close. Contactor $4A$ is then connected to relay $2AX$, and that relay governs the closing of $4A$.

Two accelerating relays suffice to control any number of accelerating contactors. They function alternately, and odd-numbered contactors are controlled by relay $1AX$, whereas even-numbered contactors are controlled by relay $2AX$. One precaution must be observed: it is essential that the electrical interlocks of the accelerating contactors be so adjusted that the normally open contacts close slightly before the normally closed contacts open, and this adjustment must be maintained. A refinement which is not indicated in Fig. 10-3, but may be found on actual controllers, is the addition of a small holding coil in series with the accelerating relay contact which insures good contact force and safeguards against possible shattering of the relay contact.

SECONDARY-FREQUENCY ACCELERATION

While a wound-rotor motor accelerates to full speed, its secondary voltage drops from standstill voltage to nearly zero. Correspondingly, its secondary frequency drops from line frequency to nearly zero, voltage and frequency being proportional to the slip. Thus, by devising relays which respond to rotor voltage and frequency condition, the closing of accelerating contactors can be made a function of motor speed.

In Fig. 11-3, accelerating relays $1AX$, $2AX$, $3AX$, $4AX$ are con-

SECONDARY-FREQUENCY ACCELERATION

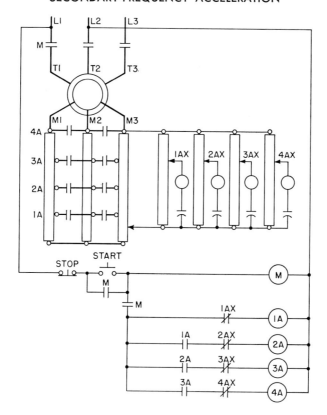

Fig. 11-3 Starting control circuit with secondary frequency-responsive accelerating relays. M = line contactor; $1A, 2A, 3A, 4A$ = accelerating contactors; $1AX, 2AX, 3AX, 4AX$ = accelerating relays.

nected in the secondary circuit. Each relay coil is connected through a capacitor to a potentiometer resistor, which in turn is connected across one leg of the starting resistor. The inductance of each relay coil, in combination with the capacitor, forms a series-resonant circuit so that a high current flows within a certain frequency range, the band width of the frequency range being determined by the adjustment of the resistor. When the frequency drops below the resonant frequency range, the relay coil current drops as a function of frequency, the actual current being adjustable by shifting the potentiometer resistor tap. With the motor deenergized, the accelerating relays are dropped out. When the start button is pressed and line contactor M closes, line frequency is applied to the accelerating relay coils, a high current flows

102 CONTROL OF WOUND-ROTOR MOTORS

in the relay coils, and all relays open their contacts immediately. As the motor accelerates, the relay coil currents drop and, by proper adjustment, relays $1AX$, $2AX$, $3AX$, $4AX$ drop out at predetermined secondary frequencies. That is to say, these relays close at predetermined motor speeds, causing the accelerating contactors to close in sequence.

DEFINITE-TIME ACCELERATION

Current-limit and secondary-frequency control of acceleration determine the instant at which subsequent accelerating contactors close by measuring some operating quantity of the motor. The contactor-closing sequence adjusts itself automatically to varying load conditions and maintains uniform accelerating current and torque peaks. This is true whether machinery is started light or loaded. When the load is light, the accelerating time is short; when the load is heavy, the accelerating time is longer. However, there is one requirement which must be fulfilled to obtain closing of the accelerating contactors: the motor must start on the first step, that is, it must start when the line contactor closes and all starting resistance is inserted in the secondary circuit. Should the motor fail to start, the primary current will not drop, and likewise the secondary frequency will not drop. The accelerating contactors will be prevented from closing and the motor will stall. The starting resistor is thus in danger of burning out. Even if the motor should start but fail to accelerate to a speed permitting the current-limit or secondary-frequency relays to close, the accelerating contactors will not close.

Many machines operate in widely varying ambient temperatures. On a warm summer day, when grease and oil are soft and fluid, the torque required to start a machine is much lower than it is on a cold winter morning, especially after a week-end shutdown, when grease and oil have congealed and friction is high. With current-limit or secondary-frequency acceleration, the relays have to be adjusted to meet the starting-current peaks necessitated by the most adverse condition. Particularly if the motor is started frequently, repeated high current peaks will increase maintenance costs and shorten the life of the motor.

Definite-time control of acceleration overcomes this disadvantage. With this system, the closing of contactors is made a function of time, and the intervals between contactors closings are adjusted so as to obtain smooth acceleration and uniform peaks under average load conditions. If a particularly heavy starting condition occurs and

DEFINITE-TIME ACCELERATION 103

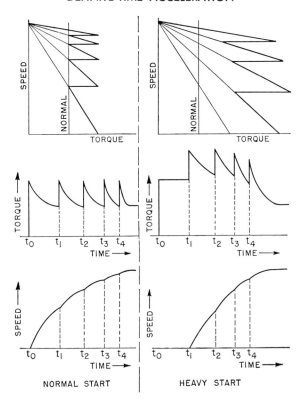

Fig. 12-3 Comparison between normal and heavy start with definite-time acceleration.

the motor fails to accelerate on the first starter step, the first accelerating contactor closes after a definite time delay, increasing the starting torque. If the motor still fails to start, the second accelerating contactor closes and increases the starting torque still further, so that the motor has a chance to break static friction and start the machine. Under such conditions, the accelerating current peaks are higher than normal, but they occur only occasionally, whereas on normal starts the peaks are lower. Definite-time acceleration is illustrated in Fig. 12-3. Diagrams on the left side, similar to those in Fig. 6-3, illustrate a normal start; corresponding diagrams on the right illustrate a heavy start. In both cases, the same starter, resistor layout, and time intervals between accelerating contactors closings are used.

There are three basic methods of obtaining time-delay intervals be-

CONTROL OF WOUND-ROTOR MOTORS

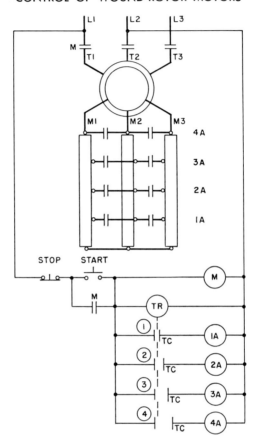

Fig. 13-3 Starting control circuit with definite-time acceleration by mechanical timer. M = line contactor; $1A$, $2A$, $3A$, $4A$ = accelerating contactors; TR = definite-time relay.

tween accelerating contactors: by the use of mechanical timers, definite time-sequence interlocks, or individual timing relays. Shown in Fig. 13-3 are the essential circuits of a starter with definite-time acceleration by means of a mechanical timer. TR is a timing relay, either solenoid- or motor-operated. This relay contains a multiplicity of contacts, as many as there are accelerating contactors. These contacts are usually arranged on a common shaft so that they close in time sequence, with adjustable time intervals between them.

When the start pushbutton is pressed, line contactor M closes and seals itself in. At the same time relay TR is energized. Following a time delay, contact 1 closes, causing contactor 1A to close. After an additional time delay, contact 2 of relay TR closes, which energizes contactor 2A. Contacts 3 and 4 of relay TR close after further time delays, causing contactors 3A and 4A to close, which completes the accelerating sequence.

An inherent disadvantage of mechanical timers is their comparatively short life, which is only a fraction of the normal life expectancy of most contactors. Longer life can be obtained by the use of time-delay interlocks, such as pneumatic timing interlocks. Principal control circuits of a starter using time-delay interlocks for sequencing the accelerating contactor closing are illustrated in Fig. 14-3.

Pressing the start button causes line contactor M to close and seal

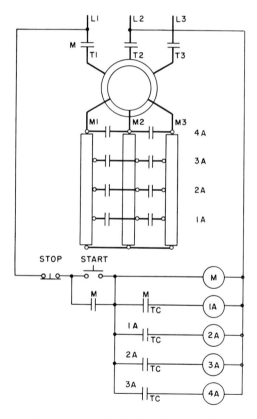

Fig. 14-3 Starting control circuit with definite-time interlocks. M = line contactor; $1A, 2A, 3A, 4A$ = accelerating contactors.

CONTROL OF WOUND-ROTOR MOTORS

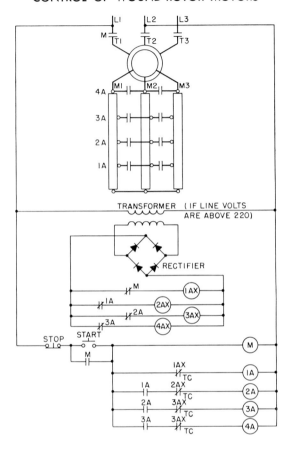

Fig. 15-3 Starting control circuit with acceleration by definite-time relays. M = line contactor; $1A, 2A, 3A, 4A$ = accelerating contactors; $1AX, 2AX, 3AX, 4AX$ = accelerating relays.

itself in. A definite-time interlock on M closes after a time interval, energizing accelerating contactor $1A$. A definite-time interlock on $1A$ closes with a time delay and energizes $2A$. In a similar manner, definite-time interlocks on $2A$ and $3A$ close with time delays and energize accelerating contactors $3A$ and $4A$. This completes the starting cycle.

Definite-time interlocks, because of their inherently longer life, are suitable for use on motors which are started frequently. Still better life can be obtained with magnetic flux-decay relays. Such relays have found considerable application on heavy-duty intermittent material-

DEFINITE-TIME ACCELERATION 107

handling drives such as cranes, mine hoists, car dumpers, steel mill auxiliaries, and the like. A supply of direct current must be provided for the relay coils. In Fig. 15-3, the accelerating relay coils are connected to a semiconductor rectifier, which is connected to the control source directly or through a transformer if the control voltage is higher than 220 volts. The rectifier can be omitted if a separate d-c control supply is available for operating the relays.

With the motor at standstill, all accelerating relays are energized, and their contacts are open. When M closes upon pressing the start pushbutton, coil $1AX$ is deenergized through a normally closed interlock on M. Relay $1AX$ drops out and closes its contact with a time delay, causing contactor $1A$ to close. Coil $2AX$ is then deenergized, relay $2AX$ drops out with a time delay and causes contactor $2A$ to close. In a similar manner, contactors $3A$ and $4A$ close with time intervals between them, as governed by relays $3AX$ and $4AX$.

In the preceding circuit diagrams, the secondary resistor has been shown as consisting of three legs, wye-connected. Secondary contactors are so selected that they are capable of carrying the full secondary current of the motor. Large motors may have secondary currents of the order of several hundred amperes, and secondary contactors capable of carrying full secondary current may be rather bulky and expensive. Contactor size may be reduced considerably by arranging the resistor so that several parallel paths are provided.

For intermediate-size motors, the double-wye arrangement in accordance with Fig. 16-3 may be used. The starting resistor consists

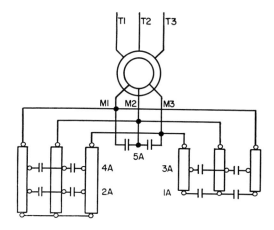

Fig. 16-3 Double-wye connection of secondary resistor.

CONTROL OF WOUND-ROTOR MOTORS

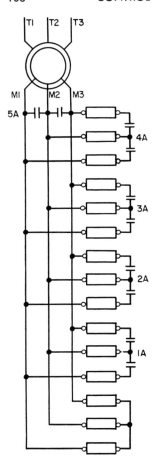

Fig. 17-3 Multiple-wye connection of secondary resistor.

of two parallel legs, each wye-connected. Accelerating contactors short-circuit sections in alternate legs. For very large motors having secondary currents of the order of 1000 amperes or larger, the multiple-wye connection may be employed as illustrated in Fig. 17-3. Resistor sections are arranged in several legs, which are progressively paralleled by accelerating contactors. Closing of the secondary contactors may be controlled by either of the previously described starting-control methods.

REVERSING

Wound-rotor motors are reversed in the same manner as squirrel-cage motors, that is, by interchanging two phases of the primary winding. In magnetic controllers, two contactors are required for interchanging the stator connections, and the same basic reversing circuits are used as for squirrel-cage motors. With a reversing controller performance is possible in all four quadrants. Figure 18-3 illustrates the performance of a wound-rotor motor in all four quadrants with various amounts of per unit resistance in the secondary circuit.

With no resistance in the secondary circuit, wound-rotor motor performance is very similar to that of a squirrel-cage motor as illustrated in Fig. 20-2. Motoring performance in both directions of rotation is obtained in quadrants 1 and 3. When the motor is driven by a load in the same direction of rotation as its flux, the motor runs at a speed above synchronous and acts as an induction generator, feeding power into the line. Dynamic braking performance is obtained in quadrants 2 and 4. This performance is also called by some authors "regenerative braking." With the slip rings short-circuited, the speed rises slightly above synchronous as long as maximum torque is not exceeded. If resistance is inserted in the rotor circuit, the dynamic braking speed rises above synchronous. At unity overhauling torque the rise in speed above synchronous is equal in per unit to the total secondary resistance in per unit. As during motoring performance, dynamic braking speed varies directly proportional to the overhauling torque on the motor shaft. Dynamic braking at speeds above synchronous can be used, as for example in lowering a load on a crane hook under the influence of gravity, as long as the secondary resistance is kept sufficiently low to prevent excessive motor speeds.

A wound-rotor motor can be plugged by reversing two stator leads, just like a squirrel-cage motor. In Fig. 18-3 plugging performance is indicated in quadrants 2 and 4. The motor then turns in a direction opposite to rotation of flux. The load is restrained by the countertorque which the motor develops. With the slip rings short-circuited, the plugging curve is similar to that of a squirrel-cage motor. It is impracticable to plug the wound-rotor motor with no resistance in the secondary circuit, because the resultant high rotor current would damage the slip rings and brushes.

Inserting resistance in the rotor circuit materially alters the plugging curves. With a secondary resistance of unity or more, the speed-torque curves become straight lines in the plugging quadrant. The plugging torque rises with rising speed. This means it is possible to lower a load, that is under the influence of gravity, by countertorque. Stable operation results as long as sufficient secondary resistance is used so that plugging torque increases with speed.

CONTROL OF WOUND-ROTOR MOTORS

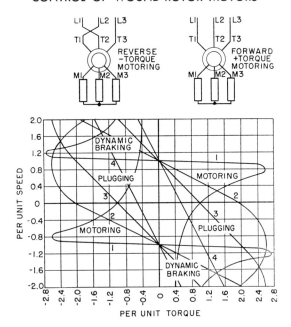

Fig. 18-3 Typical motoring, plugging, and dynamic braking curves of a wound-rotor induction motor. Curves 1 = basic motor curves; curves 2 = secondary resistance 0.5; curves 3 = secondary resistance 1.0; curves 4 = secondary resistance 2.0.

By proper selection of secondary resistance, it is possible to adjust plugging torque and plugging current to suit the requirements of the drive. At negative synchronous speed, the plugging torque and the corresponding secondary current is twice the starting torque which would be obtained with a given amount of secondary resistance. If the motor is running at full speed the torque developed at the instant of plugging in per unit is approximately

$$T_p = \frac{2}{2R_i + R_s} \quad (26) \blacksquare$$

where T_p = per unit motor torque
R_i = per unit internal rotor resistance per leg
R_s = per unit resistance per leg in series with rotor

On many reversing plugging controllers, a secondary resistance of approximately 2.0 per unit is provided which obtains a starting torque of 0.5 and an initial plugging torque of unity.

WOUND-ROTOR MOTOR STARTERS

Wound-rotor motors are started with resistance connected in the secondary circuit. Starters may be nonreversing or reversing. A standard general-purpose starter includes the following:

1. Line contactor or pair of reversing contactors
2. A set of accelerating contactors
3. A set of accelerating relays
4. Thermal overload relays
5. Starting resistor
6. Control fuses and control transformer as optional items

Line or reversing contactors are selected for the same horsepower rating as contactors for full-voltage starters of squirrel-cage motors. Final accelerating contactors are selected for carrying the actual motor secondary current. If two-pole contactors are used open-delta-connected, the motor secondary current should not exceed the continuous current rating of the contactor. If triple-pole contactors are used delta-connected, they are applied so that motor secondary current does not exceed 150 per cent of the continuous current rating of the contactor. Intermediate accelerating contactors should have a current rating of not less than one sixth the maximum current peak they have to carry. It is customary to make intermediate accelerating contactors one size smaller than the final accelerating contactor. Since the accelerating contactors do not interrupt current, they are not equipped with blowouts.

NEMA has established a standard number of accelerating contactors to be used. They are listed in Table 4, which applies to general-purpose starters.

Table 4. Number of Accelerating Contactors Used on Wound-Rotor Motor Starters

Maximum Horsepower	Minimum Number of Accelerating Contactors
15	1
75	2
150	3
300	4
600	5
1200	6

CONTROL OF WOUND-ROTOR MOTORS

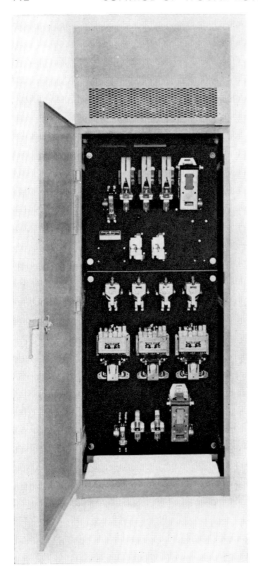

Fig. 19-3 Low-voltage starter for wound-rotor induction motor.

A typical floor-mounted enclosed starter is shown in Fig. 19-3. The door is open to show the devices on the panel. Beginning at the top, there are the line contactor, two overload relays, four accelerating relays, three intermediate accelerating contactors, and the final accelerating contactor. The starting resistor is mounted in a ventilated enclosure on top of the panel.

WOUND-ROTOR MOTOR STARTERS

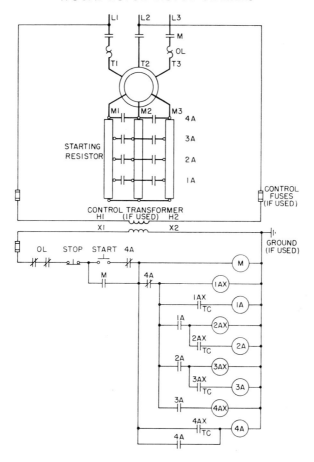

Fig. 20-3 Definite-time starter for wound-rotor induction motor. M = line contactor; $1A$, $2A$, $3A$, $4A$ = accelerating contactors; $1AX$, $2AX$, $3AX$, $4AX$ = accelerating relays; OL = overload relays.

Elementary connections of a five-point wound-rotor motor starter with definite-time acceleration are shown in Fig. 20-3. This starter contains four accelerating contactors, and the control connections indicate how control fuses and a control transformer are connected in the circuit. To be sure that, when the motor is started, all accelerating contactors are open, the pick-up circuit of line contactor M is established through a normally closed interlock on the last accelerating contactor $4A$. After the line contactor has closed, accelerating relays $1AX$, $2AX$, $3AX$, and $4AX$ are energized in sequence. These accelerating relays are of the time-delay closing type and each relay energizes

114   CONTROL OF WOUND-ROTOR MOTORS

its associated accelerating contactor with a time delay. Each contactor energizes the accelerating relay associated with the next accelerating contactor, thus causing the accelerating contactors to close with time intervals between them. When the last accelerating contactor $4A$ has closed, a normally closed interlock on that contactor opens the coil circuit to the intermediate accelerating contactors and the accelerating relays which then drop out. The final accelerating contactor seals itself in. Deenergizing the coils of the intermediate accelerating contactors and the accelerating relays eliminates coil heating and reduces heat losses and associated temperature rise within the control enclosure.

CRANE SERVICE

An important field of application for wound-rotor motors are cranes and similar material-handling drives. Many motors used on such applications are not called upon to deliver their rated load continuously, but they are all started and stopped frequently. These motors are run intermittently; that is, running periods alternate with standstill periods. Motors used for this type of service are applied according to their intermittent rating. Such ratings may be established on a 15-, 30-, or 60-minute basis, depending on the severity of the service. Selection of motors is often governed by maximum torque and maximum starting torque which the motor is able to deliver, rather than by heating limitations under continuous load conditions. However, if the crane is in continuous operation, as is frequently the case for grab-bucket or scrap-handling magnet cranes, the thermal capacity of the motor has to be analyzed on a repetitive duty-cycle basis.

Controllers for cranes and other material-handling equipments are designed according to the severity of the service to which the equipment is subjected. To assist in selecting proper control equipment, NEMA has established five standard crane classifications which are used as a criterion for choosing and rating components entering into the controller design.

Class I. Stand-by service designates cranes which are used primarily during the erection period to place machinery, and thereafter are used only as stand-bys for maintenance purposes in servicing the machinery. Stand-by service conditions are encountered in power plants, substations, pumping stations, minor freight stations, and the like. Although the equipment is used only infrequently after the original erection period, a high degree of reliability is required, and the service may be quite severe while the equipment is in actual use.

CRANE CONTROL

Class II. Light industrial service applies to cranes which are not used frequently and which, during a large portion of their use time, handle less than their rated load. Light industrial service conditions are encountered in warehouses, in assembly areas of manufacturing plants, on side floors of foundries, in freight transfer stations, and the like.

Class III. Regular industrial service describes cranes which are used in industrial plants where crane service is an important part of production processes. Regular industrial service conditions are encountered in machine shops, foundries, railroad shops, and the like.

Class IV. Continuous-handling service pertains to cranes which are in operation continuously during a working day. Such cranes carry a magnet or a bucket for handling such materials as pig iron and scrap, cement, crushed stone, sand, ore, coal, and the like. Motors are selected and control equipment is designed to meet a specific duty cycle.

Class V. Steel-mill service applies to cranes which are built for specific purposes in steel mills, according to specifications issued by the individual steel mill. Such cranes must be rugged and capable of withstanding severe service, with frequent heavy loads on the motor.

Cranes built for service according to the above classifications may differ greatly in mechanical detail. However, as far as the duty of the electrical equipment is concerned, there is no great difference between Classes I, III, IV, and V. Controllers built to the same specification would perform adequately on cranes of either one of those four classes. In the interest of standardization and economy, NEMA recognizes one type of control equipment as being suitable for these classifications. However, Class II equipments are subjected to a significantly lighter duty, hence NEMA standards recognize a lighter type of control as adequate for Class II cranes.

CRANE CONTROL

Controllers for cranes and similar material-handling equipments are manually operated, generally by master switches, sometimes by pushbuttons. The operator must be able to start and stop the motor and to vary its rate of acceleration and retardation, and its steady-state speed at will. A complete controller consists of magnetic control panel, resistor, master switch or pushbutton station, and limit switches, if required.

For contactors to be used on cranes and other intermittent-duty drives, NEMA has established ampere and horsepower ratings, which are listed in Table 5. Smaller contactors, the thermal ability of which is

116 CONTROL OF WOUND-ROTOR MOTORS

limited, are applied to intermittent drives on the same basis as to continuous drives. Larger contactors have a considerable heat-storing ability, and higher ampere and horsepower ratings are permitted when these motors are used with intermittently rated motors.

NEMA has also standardized the number of accelerating contactors to be used on crane control panels. They are listed in Table 6. Any low-torque or plugging contactors which may be used on a controller are in addition to the accelerating contactors listed in the table.

Crane controllers can be divided in two major categories, according to the type of circuit employed. One category is the reversing drives used on trolleys and bridges. On such drives the load consists of a combination friction and inertia load which is accelerated and retarded in both directions of travel. The motor drives the load in both directions of rotation in quadrants 1 and 3. In order to bring the drive to a stop, retardation, generally by countertorque, is applied and the motor operates then in quadrants 2 and 4.

The second category comprises controllers for hoists. The peculiar operating condition which has to be met by these controllers is the lowering of a load which is under the pull of gravity. When the load is hoisted, the motor delivers torque to the load. Speed-torque curves are in quadrant 1. An empty hook and some light load are usually not heavy enough to overcome the friction of the drive; that is, the motor is not overhauled, and a positive driving torque is delivered by the motor to lower the load. Operation is in quadrant 3. When a load on a hook or an empty or loaded bucket is lowered, the load overhauls the drive; that is, the motor absorbs power from the load, or a load brake restrains the load. When the motor absorbs power from the load, operation is in quadrant 4. If a load brake restrains the load, the motor delivers a driving torque and operation is in quadrant 3.

Some authors present speed-load curves to describe crane performance. Such curves give speed as a function of load on the crane. Mechanical efficiency enters into the determination of load. To the control engineer, it is more convenient to present speed as a function of motor torque, since performance calculations must be based on torque delivered or absorbed at the motor shaft. Speed-torque curves are used in this chapter to describe the performance of crane controllers.

NEMA has established standards governing the components to be included in controllers for the various crane classes. These standards will not be described in detail; those interested in the subject should refer to NEMA Standards for Industrial Control, Parts 42, 43, and 44.

Cranes are operated either by master switches installed in the cab

CRANE CONTROL

Table 5. Standard NEMA Contactor Ratings for Use with Wound-Rotor Motors in A-C Crane Service

Size	Open Continuous Rating (amperes)	Amperes	Crane Rating	
			Horsepower at	
			220 volts	440–550 volts
0	18	18	3	3
1	30	30	7½	7½
2	50	67	20	40
3	100	133	40	75
4	150	200	60	125
5	300	400	150	300
6	600	800	300	600
7	900	1200	450	900
8	1350	1800	600	1200

Table 6. Number of Accelerating Contactors Used on A-C Crane-Control Panels

Motor Horsepower Rating	Minimum Number of Accelerating Contactors *
Classes I, III, IV, V Cranes:	
15 hp or less	2
16 to 75 hp	3
76 to 200 hp	4
Above 200 hp	5
Class II Cranes:	
7½ or less	1
8 to 30	2
31 to 75	3
76 to 125	4

* Exclusive of plugging and low-torque contactors.

of the crane or from the floor by pendant pushbutton stations or rope-operated drum switches. Crane operators may be shifted from job to job; hence it is important to maintain uniformity of master-switch handle throw and corresponding motion of the crane. Among builders and users of cranes, it has become accepted practice that movement of the master-switch handle *away* from the operator corresponds to forward motion of a bridge or trolley and lowering of a hoist, whereas movement of the switch handle *toward* the operator corresponds to reverse travel or hoisting. Since many crane controllers, particularly those for hoists, require unsymmetrical contact closing sequences of the master switch in the two directions, NEMA has established a standard for correlating rotation of the master-switch shaft and the desired control operation. Standard master-switch handle throw is indicated by Fig. 21-3. In planning the assembly of master switches in a crane cab, it may be found that mounting conditions are such as to require a deviation from the standard master-switch rotation. Such conditions should be specifically stated by purchasers of crane controllers, since manufacturers, unless instructed otherwise, supply master switches with standard handle throw. On larger cranes, it is customary to provide an individual control panel for each motion. On smaller cranes all motions may be combined on one control panel. Each crane equipment includes a main line disconnect switch with provisions for locking the switch in the open position. The disconnect may be a switch in the feeder to the crane, or it may be a main line switch on the crane control panel. This important safety feature, which is called for in various safety codes, makes it unnecessary for an electrician working on a crane to depend on somebody away from the crane to interrupt power. On the larger cranes, a dis-

Fig. 21-3 Standard handle throw of crane master switches.

connect switch may be installed on the control panel for each motion, thus permitting each motion to be energized individually for testing the control.

Overload protection is included, generally by overcurrent relays. Instantaneous overcurrent relays obtain adequate protection. However, since some types of overcurrent relays may be so fast as to trip within a half cycle, nuisance tripping may occur as a result of the magnetizing current inrush when the motor is energized. For this reason, it is customary to introduce a slight time delay of a few cycles in the overcurrent relays.

Particularly on larger cranes, the thermal capabilities of the motors are not limiting, but rather current and torque peaks which may damage the slip rings or endanger mechanical parts. Overload protection is therefore regarded as protection against excessive current and torque peaks. It is customary to set the overcurrent relays at approximately 2.5 times rated motor full-load current. Crane controllers also include undervoltage protection, requiring the operator to return his master switch to the neutral position since inadvertent starting of the crane upon return of voltage would certainly represent a major hazard.

These safety features may be included on each individual crane control panel, or they may be included on a crane protective panel mounted in the operator's cab and serving all motions of the crane.

REVERSING HOIST AND TRAVEL CONTROLLERS

Straight reversing controllers with plugging are used on bridge and trolley motions. Customarily, these controllers include a plugging feature for the purpose of retarding the drive at the ends of the forward and reverse travel. Reversing controllers are also used on hoists which are equipped with mechanical load brakes. On such hoists, a mechanical brake holds the load so that it is not accelerated in the downward direction by gravity. A driving torque is required of the motor to lower a load on the hoist. The plugging feature is omitted on hoists.

Elementary connections of a typical reversing plugging controller are shown in Fig. 22-3. Symmetrical operation is obtained in both directions of travel. The motor is under the control of a five-point master switch. Turning the master switch in the forward or reverse direction energizes the forward or reverse contactor connecting the motor to the line in either direction of rotation. As the master switch is turned step by step out of the neutral position, the secondary contactors close in sequence. If the master switch is turned rapidly to the

CONTROL OF WOUND-ROTOR MOTORS

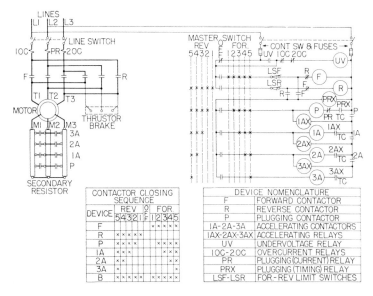

Fig. 22-3 Elementary connections of a reversing plugging a-c crane controller.

last point, the closing of the secondary contactors is governed by definite-time accelerating relays which introduce time intervals between the closing of successive contactors. Limit switches for overtravel protection are connected in the coil circuits of the reversing contactors. A Thrustor brake is connected to the primary terminals of the motor. On machinery designed to move a heavy mass, gradually setting brakes are advantageous to bring about smooth stopping. When quicker brake setting is desired, solenoid brakes are used.

Retardation of the moving masses can be accomplished by plugging. The countertorque of the motor is utilized to bring the drive to a dead stop by electric braking, or to effect rapid reversal. A current-type plugging relay PR opens its contact, as a result of the high current peak incident to the reversing of the motor stator connections, and prevents the closing of plugging contactor P. When the motor current has dropped to a low value, indicating that the motor has decelerated to a low speed, relay PR drops out and permits the closing of contactor P. A timing relay PRX introduces a slight time delay between the closing of the reversing contactors and the energizing of the coil circuit of P, in order to permit relay PR to pick up when the master switch is moved rapidly through the first position. As an alternative, motor

REVERSING HOIST AND TRAVEL CONTROLLERS 121

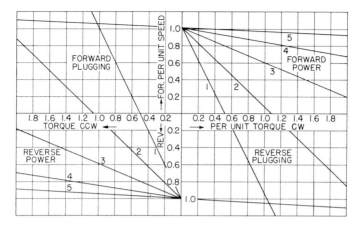

Fig. 23-3 Speed-torque curves of a reversing plugging a-c crane controller.

secondary voltage, as an indication of motor speed, is sometimes used for actuating a plugging relay.

The speed-torque curves obtained on the various master-switch positions are plotted in Fig. 23-3. The total resistance contained in the resistor, plus internal rotor resistance, is twice normal ohms. The starting torque on the first master-switch position is one half normal torque. The initial plugging torque obtained with such a resistor layout is approximately equal to normal torque. NEMA Class 152 resistors are used with Class II cranes, whereas Class I, III, IV, and V cranes use Class 162 resistors.

Undervoltage protection is obtained by undervoltage relay UV. If voltage fails, the master switch has to be returned to the off position before UV can be energized again, thus preventing inadvertent starting of the crane upon reappearance of power. Overload protection is provided by two overcurrent relays. The line switch, undervoltage relay, and overload relays may be included on the individual control panel, or these functions may be provided by the crane protective panel.

On hoist controllers with load brakes, the plugging feature is omitted. There are no relays PR and PRX. The plugging contactor is used as a low-torque contactor. Resistor layout is the same as for plugging controllers. The low starting torque on the first master-switch point is used for maneuvering a light load or an empty hook. Speed-torque curves of a hoist controller are like those shown in Fig. 23-3.

CONTROL OF WOUND-ROTOR MOTORS

Fig. 24-3 Typical reversing plugging a-c crane control panel.

A typical reversing plugging a-c crane control panel is shown in Fig. 24-3. All devices are mounted on a steel base and all wiring and device terminals are accessible from the front, since space for installing and servicing control panels on a crane is at a premium.

DUPLEX CONTROLLERS

When two motors are geared to the same drive and are controlled simultaneously from one master switch, a single duplex control panel is used. Figure 25-3 is an elementary power diagram of a duplex reversing controller. The stators of both motors are connected to the power supply through a common pair of reversing contactors. Each motor is protected individually by its own overload relays. In the diagram a common line switch is shown, assuming that in case of failure

DUPLEX CONTROLLERS

of one motor it is not desired to maintain emergency operation with a single motor. Single-motor operation could be provided by connecting an individual line switch in each motor primary circuit.

The secondary circuits of both motors are electrically separated, and each motor has its own resistor, to avoid the possibility of circulating current flowing between the two motors. However, a single set of four-pole accelerating contactors short-circuit corresponding sections of both resistors. Control-panel space is reduced thereby, as compared with a panel mounting two separate sets of contactors. Also, the use of four-pole accelerating contactors insures that resistance sections are short-circuited simultaneously in the two motor secondaries, resulting in smooth acceleration.

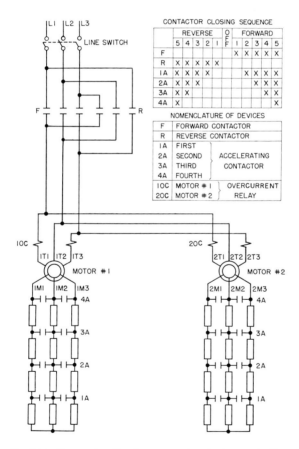

Fig. 25-3 Power circuits of a-c reversing duplex controller.

Any of the conventional methods of controlling accelerating contactor closing sequence may be used. With definite-time control of acceleration, there is no connection between power circuits and accelerating relays. If single-motor operation is desired, the faulty motor is disconnected and no changes are required in the control circuits. If current-limit relays or rotor-frequency relays are used, such relays are connected to one of the two motors. In case of single-motor operation, a transfer switch is included to connect the relays to either motor circuit.

HOIST CONTROLLERS

Alternating-current induction motors are inherently less flexible in their speed-torque characteristics than d-c motors, particularly when the motor has to develop dynamic-braking torque to retard overhauling loads. When an induction motor is driven as a generator by an overhauling load in the same direction of rotation as its flux, the motor can develop a retarding torque only at speeds above synchronous. In that speed range, control is possible by varying the resistance connected in the rotor circuit. Below synchronous speed, no speed control is possible except with countertorque obtained by plugging. More desirable braking-torque characteristics at speeds below synchronous can be obtained with controllers which are somewhat more complex than reversing plugging controllers.

When induction motors were first applied to hoists, they were used on balanced hoists, particularly mine hoists, where a loaded skip is hoisted while an empty skip is lowered. This operation requires the motor to deliver driving torque in both directions of rotation. Unbalanced hoists, especially crane hoists, used the motor only for hoisting and loads were lowered under the control of mechanical brakes, with the motor disconnected from the line. Mechanical load brakes were then introduced; these restrain the load in the lowering direction and require power from the motor to drive the load down. Because of mechanical complications and high maintenance cost, mechanical load brakes are today applied only to infrequently operated cranes.

For many years, d-c motors were used almost exclusively on cranes requiring accurate spotting of the load, and series motors are still unsurpassed for their simplicity and flexibility of control. In areas where large blocks of d-c power are available, as in steel mills, d-c motors are preferred. Cranes which are remote from power sources and which carry their own gasoline or diesel power plant are often equipped with d-c drives. Direct-current shunt motors with adjustable-voltage speed

HOIST CONTROLLERS

control are sometimes used on high-speed crane hoists requiring accurate positioning of the load. The adjustable-voltage power source is obtained by conversion equipment from an alternating-current power supply or from a generator driven by a prime mover.

From an economic point of view, the use of a-c motors for crane drives is desirable because conversion equipment, higher d-c distribution system costs, and the additional losses of the d-c distribution system can be saved. Considerable work has been done in recent years to develop control systems obtaining speed-torque curves for lowering overhauling loads which, to a varying degree, approach the performance possible with d-c motors. Such dynamic-braking a-c hoist controllers are more complex than conventional reversing controllers with speed control by changing secondary resistance. The cost of such controllers varies considerably, depending on the refinements included in the controller and the degree of performance obtainable in the lowering direction. The selection of a crane-hoist controller for a particular application is governed by economic as well as technical considerations.

Depending on the accuracy with which a load has to be handled and the flexibility required of the hoist controller, cranes can be classified according to two main categories:

1. Cranes handling material in bulk, such as grab-bucket cranes, cableways, coal and ore bridges. Loads vary between definite limits, the maximum load being determined by the dimensions of the bucket and the weight of the material handled, the minimum load being an empty bucket. Accurate inching is not required. A low-torque point is necessary to pay out cable while the bucket rests on the pile, in order to maneuver the bucket.

2. Cranes handling miscellaneous work. In this category are shop, powerhouse, and shipyard cranes, and many of the cranes used around steel mills. Loads vary widely. Accurate spotting of loads and control of lowering speeds below synchronous are essential. It is desirable that changes of speed with variations in load on any master-switch position be kept at a minimum.

A number of a-c crane-hoist control systems for wound-rotor induction motors depend on changes in motor secondary resistance and changes in motor primary connections for controlling torque and speed of the motor. The following basic control systems have found wide application:

CONTROL OF WOUND-ROTOR MOTORS

Lowering by plugging or countertorque
Lowering by dynamic braking with single-phase stator connection
Lowering by dynamic braking with unbalanced stator connection
Lowering by d-c dynamic braking

The latest development in the refining of wound-rotor motor controllers is the use of saturable reactors or transformers in the motor primary and secondary circuits. Such controllers are among the most complex and most expensive ones for use with a-c motors but, particularly when used in conjunction with regulators, they obtain performance which resembles most closely the performance of d-c motors in regard to flexibility and accuracy of speed control. However, one feature which the a-c motor drive cannot duplicate is the high light-load hoisting speed obtainable with d-c motors.

GRADUATED PLUGGING CONTROL

The oldest types of a-c hoist controllers, widely used on bulk-handling grab-bucket cranes, accomplish control of the load during lowering by connecting the motor primary circuit in the hoisting direction and inserting sufficient resistance in the secondary circuit so that the load torque is greater than the motor torque at standstill. The energy transmitted by the load to the motor is dissipated in the secondary resistor. Speed control is obtained by varying the amount of secondary resistance. Since several master-switch positions are available to control the amount of plugging or countertorque developed by the motor, such controllers are called "graduated plugging controllers."

Figure 26-3 is an elementary power diagram of a graduated plugging crane hoist controller. The speed-torque curves obtained on the various master-switch positions are plotted in Fig. 27-3. In the hoisting direction, contactor H connects the motor primary to the line with correct phase sequence. Secondary contactors are closed in sequence on the various master-switch positions. Accelerating relays, such as definite-time, current-limit, or frequency relays, govern the interval between contactors closing, independent of the speed with which the master switch is manipulated. In the hoisting direction, the controller shown does not differ from a conventional reversing controller.

When the master switch is turned in the lowering direction, nothing happens until the fifth position is reached. The motor primary is connected to the line through contactor L with reversed phase rotation, the brake is released, and all accelerating contactors close in sequence. The motor accelerates to its full speed, which, with an overhauling

GRADUATED PLUGGING CONTROL

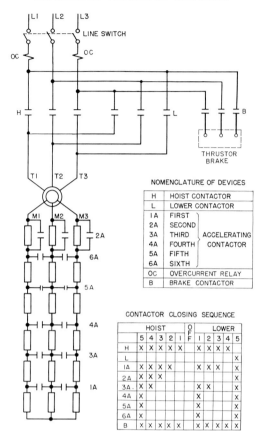

Fig. 26-3 Power circuits of graduated plugging crane hoist controller.

load, is slightly above synchronous. The load is lowered at its maximum speed in accordance with curve 5-L, a very stable curve, preventing the motor from overspeeding even with a heavy overload. The motor acts as a generator, pumping power back into the line.

To retard the load preparatory to stopping, the master switch is moved toward the off position. On the fourth position, contactor L opens and contactor H closes, connecting the motor primary to the line in the hoisting direction while the motor is still turning in the opposite direction. All accelerating contactors are dropped out, and the motor develops a weak countertorque to retard the load. Turning the master switch step by step to position 1 causes accelerating con-

CONTROL OF WOUND-ROTOR MOTORS

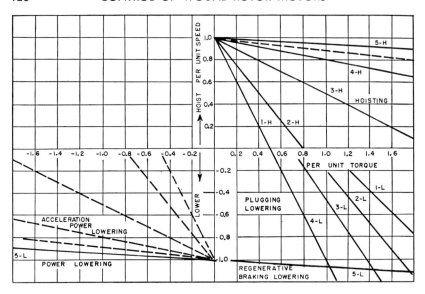

Fig. 27-3 Speed-torque curves of graduated plugging crane hoist controller.

tactors to close, and the countertorque is gradually increased, thus bringing the load to a stop preparatory to setting the brake in the off position.

Care must be taken to prevent setting of the brake during the transition between positions 4 and 5 of the master switch, while contactor L drops out and contactor H closes. A Thrustor brake, with an inherent time delay in setting, stays released during the short interval during which the motor is disconnected from the line. When a solenoid brake is used, the brake contactor is so arranged that the brake is kept energized during the transition period. As a safety feature to prevent overspeeding of the motor in the lowering direction, a voltage-sensitive relay may be connected in the secondary circuit of the motor. This relay closes secondary contactors, thereby cutting out secondary resistance as the motor approaches synchronous speed. A second voltage-sensitive relay may be used to remove power from the motor primary and to set the brake when the motor approaches standstill with the master switch in a countertorque position.

When such a controller is used on a magnet or a bucket crane, it may be desirable to jog the magnet or bucket a short distance into a pile of material in case the operator stops too soon with countertorque. To permit such jogging, a foot switch or thumb-operated auxiliary contact in the master-switch handle can be provided to obtain a light motoring torque in the lowering direction with the master switch in a slow-speed position. With this auxiliary switch, the same connections are established on the first positions of the master switch in the lowering direction as would be obtained ordinarily when the master switch is moved to the fifth position. Releasing the auxiliary switch establishes regular point 1 connections, and the motor develops a retarding countertorque.

Although plugging obtains lowering speeds between zero and synchronous, it must be recognized that the resultant speed-torque curves are not suited for all kinds of hoist applications. The curves are steep, a small variation in load corresponding to a large variation in speed. Overloads may cause overspeeding of the motor, and the operator must exercise care in manipulating the master switch. The risk of overspeeding is reduced by the use of safety relays responding to motor speed. Plugging control is well suited for bucket hoists which do not require accurate speed control and on which overloads are not encountered. The same applies to magnet cranes on which the maximum load is limited by the lifting power of the magnet. Plugging makes accurate maneuvering of loads difficult, so that this type of control is not recommended for powerhouse cranes, shipyard cranes or steelmill cranes handling hot material. However, where graduated plugging control meets operating requirements satisfactorily, it offers the advantage of simplicity. It is also less expensive and easier to maintain than other types of hoist control, since it requires a minimum of electrical equipment.

UNBALANCING OF STATOR VOLTAGE

The control systems for wound-rotor induction motors discussed so far apply balanced line voltage to the stator and manipulate resistance in the secondary circuit. Several control systems have been developed which unbalance the primary voltage applied to the motor. The resultant speed-torque curves can be used to advantage for cranes and similar material-handling equipment.

Any unbalanced polyphase system can be resolved into balanced polyphase systems of positive, negative, and zero sequence by using symmetrical components. The theory involved is not discussed here in detail; interested readers should refer to a standard textbook on the subject. When applying an unsymmetrical polyphase voltage to the stator terminals of an induction motor, zero-sequence voltage can be neglected as far as calculating motor performance is concerned, since it does not develop torque and has no effect on the speed-torque curve. Zero-sequence current, which may flow only in a delta-connected stator winding but not in a wye-connected winding, may increase the motor losses. This effect has to be considered in selecting the motor frame size for a given crane rating.

The effect of positive- and negative-sequence voltages on the speed-torque curves of an induction motor can be evaluated as follows: if the speed-torque curve of the motor is known for rated symmetrical stator voltage, and if positive- and negative-sequence voltages are known in per unit of rated motor voltage, the torque developed by either positive- or negative-sequence voltage component at any speed is equal to normal motor torque multiplied by the square of the ratio of positive- or negative-sequence voltage to normal voltage. Positive-sequence torque can thus be plotted as torque for one direction of rotation, while negative-sequence torque can be plotted for the opposite direction of rotation. Net torque of the motor is the difference between positive-sequence torque and negative-sequence torque. A single motor to which unbalanced voltage is applied can thus be visualized as a combination of two motors on a common shaft, but working against each other. One motor, having symmetrical positive-sequence voltage applied to its stator, would try to run in one direction of rotation; the other motor, having symmetrical negative-sequence voltage applied to its stator, would drive the shaft in the opposite direction of rotation. Actual shaft rotation will depend on which motor overwhelms the other.

Various methods can be used to unbalance the voltage applied to the stator artificially. Saturable reactors may be inserted in one or two of the stator phases, and the degree of unbalance can be varied by varying the d-c excitation of these reactors. Another method is the use of an autotransformer with overhanging winding across two stator phases. This method is indicated schematically in the upper sketch of Fig. 28-3. The degree of unbalance of primary voltage applied to the motor stator is varied by shifting taps on the autotransformer winding. Control of the motor is effected by suitable selection of degree of primary unbalance, as well as selection of secondary resistance. The general shape of speed-torque curves which can be obtained with this system are indicated in Fig. 28-3. These curves are plotted for a hypothetical motor having a maximum torque of 2.75. Secondary resistance is assumed to be 0.6. The dotted curves represent the speed-

UNBALANCING OF STATOR VOLTAGE

torque curves obtained for positive- and negative-sequence voltages equal to 0.85 and 0.6 line voltage. Since torque varies as the square of applied voltage, maximum torque is reduced to 2.0 and 1.0, respectively. An extreme case of unbalance is obtained when motor stator terminals $T2$ and $T3$ are connected together on some point on the autotransformer. This is equivalent to applying single-phase voltage to the stator, and positive- and negative-sequence voltages are equal. Curve A, in Fig. 28-3, is plotted on the assumption that positive- and negative-sequence voltages are both 0.85 line voltage. The resultant curves show that zero torque is developed at standstill. The motor does not start, but it is capable of retarding an overhauling load. Such

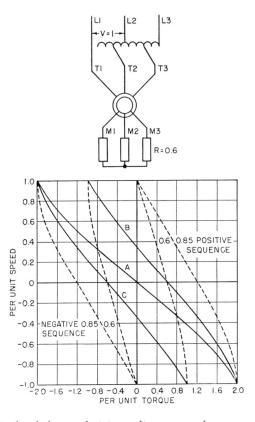

Fig. 28-3 Effect of unbalance of stator voltage on performance of wound-rotor induction motor. Curve A = positive- and negative-sequence voltages, each 0.85; curve B = positive-sequence voltage 0.85, negative-sequence voltage 0.60; curve C = positive-sequence voltage 0.60, negative-sequence voltage 0.85.

a speed-torque curve is a desirable one to use on a crane hoist, since it obtains lowering speeds between zero and synchronous. Since torque increases with increasing speed, such a curve obtains stable operation and prevents a load from attaining excessive speed, as long as the load stays within the rating of the hoist.

Curve B is plotted on the assumption that $T2$ and $T3$ are connected to the autotransformer in such a manner that positive-sequence voltage is 0.85 and negative-sequence voltage 0.6 line voltage. Curve B is the difference between dotted curves for 0.85 positive- and 0.6 negative-sequence voltage. In a similar manner, curve C is plotted on the assumption that $T2$ and $T3$ are reversed on the autotransformer, so that positive-sequence voltage is 0.6 and negative-sequence voltage 0.85 line voltage. Curve C is the difference between the dotted curves for 0.85 negative- and 0.6 positive-sequence voltage.

Curves B and C show that no-load speeds below synchronous speed can be obtained. This is an important advantage over the previously described control systems. A no-load hoisting speed below synchronous speed means that an empty hook can be hoisted slowly, or that slack cable can be taken up slowly. A no-load lowering speed of less than synchronous speed offers the advantage that a light load can be lowered by dynamic braking at less than synchronous speed which permits more accurate spotting of a load than if this load would have to be accelerated to approximately synchronous speed by a driving torque of the motor in the lowering direction.

While the speed-torque curves obtained with stator unbalancing have desirable features, one basic disadvantage of stator unbalancing must be considered. Motor losses are equal to the sum of the losses caused by the positive- and negative-sequence voltages, whereas the developed torque is equal to the difference of the torques developed by the positive- and negative-sequence voltages. This means that, for a given net output, a larger motor frame size may have to be selected. As an alternative, the motor may attain higher than normal temperature, which may reduce its service-life expectancy.

SINGLE-PHASE DYNAMIC LOWERING CONTROL

Single-phase dynamic lowering crane hoist controllers utilize the single-phasing of the motor primary to obtain subsynchronous lowering speed with an overhauling load (see curve A in Fig. 28-3). Elementary power circuits of a crane hoist controller employing a single-phase dynamic lowering connection are shown in Fig. 29-3. In the

SINGLE-PHASE DYNAMIC LOWERING CONTROL

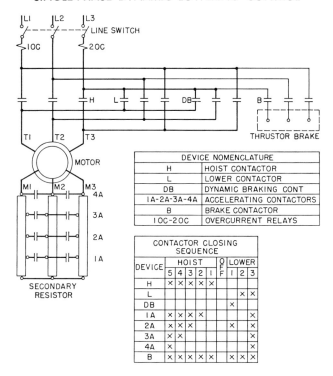

Fig. 29-3 Power circuits of crane hoist controller with single-phase dynamic lowering.

hoisting direction, contactor-closing sequence is the same as for a conventional reversing controller. That means, that the hoisting operation is performed with symmetrical line voltage applied to the motor primary. When the master switch is turned to the first position lower, contactor DB closes. Motor primary terminals $T1$ and $T2$ are tied together and connected to line phase $L3$, whereas motor primary terminal $T3$ is connected to line phase $L1$. Single-phase voltage is applied to the motor primary winding, and contactor $2A$ closes, leaving a substantial amount of resistance in the motor secondary circuit. Referring to speed-torque curves of Fig. 30-3, performance on the first point lowering is indicated by curve $1-L$. Mechanical efficiency of crane hoists is generally of the order of 0.8. Hoists with roller bearings may have efficiencies as high as 0.9, whereas poorly adjusted or worn sleeve bearings may reduce efficiency below 0.8. When the load overhauls the hoist, losses have to be supplied by the kinetic energy of the

134 CONTROL OF WOUND-ROTOR MOTORS

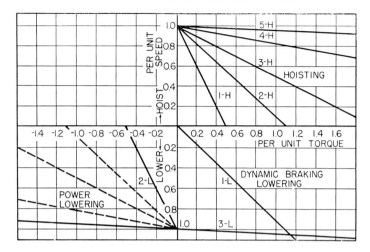

Fig. 30-3 Speed-torque curves of crane hoist controller with single-phase dynamic lowering.

load. Considering the reverse efficiency of the hoist, full load represents an overhauling torque on the motor shaft of the order of 0.65. Curve 1-L indicates that the single-phase dynamic lowering control as shown obtains a full-load lowering speed of approximately 0.6 full-load hoisting speed.

Turning the master switch to the second position lower causes contactors DB and $2A$ to open and L to close. The motor primary is then connected to the line with symmetrical three-phase voltage applied and with approximately twice normal ohms connected in the secondary circuit. The motor develops a limited power torque in the lowering direction in accordance with curve 2-L. This master-switch position is intended to be used for driving down a light load or an empty hook which would not be sufficient to overcome friction and overhaul the motor. If this master-switch position were used for lowering a heavy load, overspeeding of the hoist motor would result. To avoid this danger, relays are included which cause accelerating contactors to close and short-circuit the secondary resistor should the motor accelerate above synchronous speed.

In the third master-switch position, all accelerating contactors close in sequence under the control of accelerating relays. Automatic acceleration is indicated by the broken curves in the power lowering

UNBALANCED STATOR VOLTAGE LOWERING CONTROL   135

quadrant of Fig. 30-3. With overhauling loads, regenerative braking takes place in accordance with curve 3-L.

Single-phase lowering is accompanied by greater heating of the motor than would occur during balanced-voltage operation. During lowering, the motor is heated approximately twice as much as when it is hoisting full load. Therefore, the first master-switch position should not be used to lower the load through its full travel. The proper way of operating a single-phase lowering controller is to advance the master switch to the second or third position and to lower at full speed. As the end of the lowering travel is approached, the master switch is returned to the first position. This retards the load to a subsynchronous speed, preparatory to stopping by returning the master switch to the off position and setting the brake. Reducing the full-load lowering speed from approximately synchronous to approximately 0.7 reduces the kinetic energy stored in the moving masses, which has to be dissipated by the brake, to approximately 50 per cent of the kinetic energy stored at full speed. This results in considerable reduction of brake wear and increases the accuracy with which a load can be stopped. However, since the single-phase dynamic lowering control does not permit reduction of the full-load lowering speed to a low value, this type of control should only be used when accurate spotting of a heavy load is not required.

UNBALANCED STATOR VOLTAGE LOWERING CONTROL

Curves B and C of Fig. 28-3 indicate that, by proper selection of primary voltage conditions, dynamic braking can be obtained at speeds below synchronism. One form of a-c crane hoist controller, using unbalanced primary voltage, has been developed by Cutler-Hammer, Inc. Figure 31-3 is an elementary diagram of the power circuits. Hoisting connections are conventional, with balanced voltage applied to the stator terminals and speed controlled by short-circuiting secondary resistor sections. To reduce the amount of resistance required to obtain a low starting torque on the first master-switch position, one secondary phase is left open. Closing of contactor $4LA$ on subsequent master-switch positions balances the three secondary phases.

When the master switch is turned to the lowering positions, an autotransformer is connected to line phases $L2$ and $L3$. Stator terminal $T3$ is connected to a fixed tap on an overhung section of the transformer winding. Stator terminal $T2$ is connected to various taps on

CONTROL OF WOUND-ROTOR MOTORS

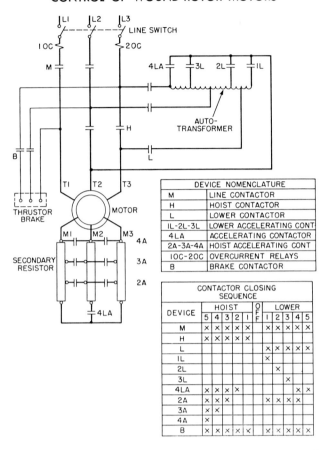

Fig. 31-3 Power circuits of crane hoist controller with unbalanced primary connections. (Courtesy Cutler-Hammer, Inc.)

the transformer winding, depending on the master-switch position. In Fig. 32-3, simplified stator connections for each master-switch position are shown, together with the corresponding voltage vector diagrams. The balanced line voltage triangle is indicated by broken lines. The solid lines represent the voltage triangles applied to the motor primary terminals. These voltages are quite unbalanced, containing various amounts of positive- and negative-sequence voltage. In master-switch positions 1 to 4, the amount of secondary resistance is fixed and the amount of voltage unbalance is varied. Thus, a fixed amount of voltage unbalance is obtained on each master-switch posi-

UNBALANCED STATOR VOLTAGE LOWERING CONTROL 137

tion. In the fifth master-switch position, additional resistance is inserted in the secondary circuit.

Speed-torque curves obtained with this type of control are plotted in Fig. 33-3. Lowering speeds above and below synchronous are obtained with overhauling loads. The slope of the curves, that is the variation of speed with torque, is considerably less than with a plugging controller, although the curves are not so flat as those obtainable with other types of hoist controllers. The first lowering curve is similar to the one obtained with single-phase dynamic lowering, and subsequent curves are spaced well apart. Power torque for lowering an empty hook is available on master-switch positions 2 to 5.

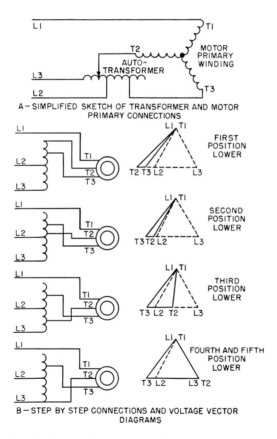

Fig. 32-3 Details of unbalanced stator lowering connections. (Courtesy Cutler-Hammer, Inc.)

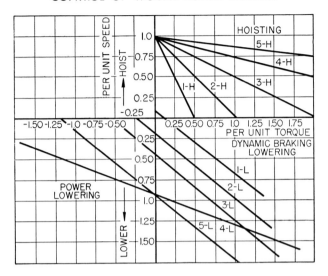

Fig. 33-3 Speed-torque curves of crane hoist controller with unbalanced primary connections. (Courtesy Cutler-Hammer, Inc.)

DIRECT-CURRENT DYNAMIC BRAKING LOWERING

If the stator of an induction motor is excited with direct current, a d-c flux is set up in the motor which is stationary in space. When an overhauling load drives the rotor, the rotor conductors intersect the d-c flux, and a dynamic braking torque is developed in the same manner as previously described for squirrel-cage motors. The speed-torque characteristics are influenced by the magnitude of the d-c exciting current, increased d-c excitation corresponding to increased braking torque or lower speed of the overhauling load. Furthermore, the speed attained at a given overhauling torque can be varied by connecting an adjustable amount of resistance in the rotor circuit. This resistance governs the drop in the secondary circuit, which in turn must be counterbalanced by the emf generated in the rotor. With d-c flux at a constant value, speed increases with increasing secondary resistance.

To plot speed-torque curves, it is necessary to obtain from the motor designer the amount of resistance required for operating the motor at synchronous speed with base torque at the motor shaft, the maximum torque, and the amount of d-c excitation. If d-c excitation is held constant, speed at a given torque varies directly in proportion to sec-

DIRECT-CURRENT DYNAMIC BRAKING LOWERING

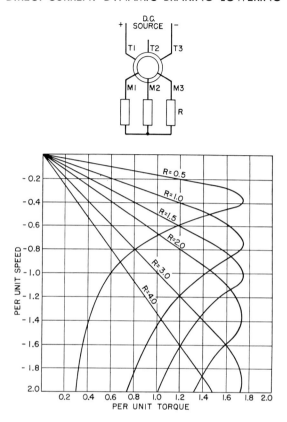

Fig. 34-3 Typical d-c dynamic braking performance curves of wound-rotor induction motor with d-c excitation of 1.75 times full-load a-c current.

ondary resistance. For instance, halving the resistance halves the speed.

For estimating purposes, the following relation can be assumed to apply to an average wound-rotor motor: If a d-c excitation equal to 1.75 times full load a-c rms current is applied to the stator phases, maximum torque is approximately 1.75. Synchronous speed at base torque is obtained with a secondary resistance of 3.0. In Fig. 34-3, speed-torque curves for this average case are plotted for various amounts of secondary resistance. These curves are typical, and individual motors may deviate from this average condition considerably. With a given amount of secondary resistance, torque increases with increasing speed, as long as maximum torque is not exceeded.

140 CONTROL OF WOUND-ROTOR MOTORS

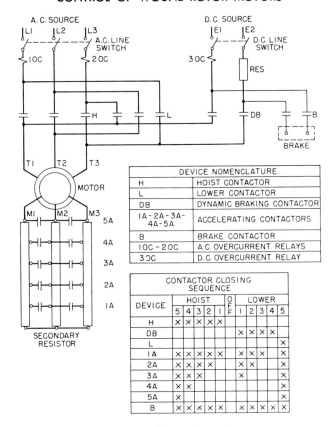

Fig. 35-3 Power circuits of crane hoist controller with d-c dynamic braking.

Figure 35-3 shows power circuits of a crane hoist controller which utilizes d-c dynamic braking for controlling the lowering speed. In addition to the a-c power source, there must be a d-c power source to supply d-c excitation of the motor during lowering. This d-c source may be a small exciter set, or it may be an available constant-voltage d-c source. Provision should be made to permit adjustment of the d-c excitation current, either by an exciter field rheostat or by an adjustable resistor in series with the constant-voltage source.

In the hoisting direction, the motor primary is connected to the a-c line through contactor H. Secondary accelerating contactors close in sequence in the various master-switch positions. Accelerating relays provide the required intervals between contactors closing when the master switch is moved rapidly to the last position.

DIRECT-CURRENT DYNAMIC BRAKING LOWERING

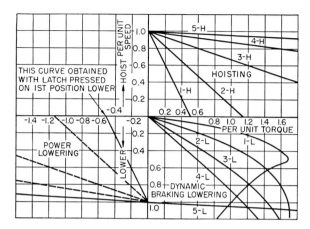

Fig. 36-3 Speed-torque curves of crane hoist controller with d-c dynamic braking.

Turning the master switch to the first four positions lowering causes contactor *DB* to close. The motor primary winding is connected to the d-c source. The farther the master switch is advanced, the more resistance is connected in the secondary circuit by dropping out accelerating contactors. The speed-torque curves are plotted in Fig. 36-3. The dynamic braking lowering curves slope away from the origin at various angles. D-c dynamic braking provides stable lowering of overhauling loads, and the speed is increased by increasing the amount of resistance in the secondary circuit. Power lowering of an empty hook is not possible with the d-c dynamic braking connection. For lowering light loads, a limited amount of power lowering torque is made available on the first two master-switch positions by pressing a latch on the master-switch handle. When this is done, contactor *DB* opens, contactors *L* and *1A* close, and the motor is connected to the a-c line with approximately twice normal ohms in the secondary circuit.

When the master switch is turned to the fifth position lowering, the motor primary is disconnected from the d-c source and connected to the line with reversed phase rotation. Accelerating contactors close in sequence. Curve 5-*L* indicates motor performance, and conventional regenerative braking is obtained with overhauling loads. When the load is retarded from high speed by returning the master switch toward the off position, the accelerating contactors first drop out and then reclose in definite time sequence. Should the motor suddenly be thrown from curve 5-*L* to curve 1-*L*, a runaway condition would exist if the

142 CONTROL OF WOUND-ROTOR MOTORS

load on the motor were higher than the point of intersection between curve 5-L and curve 1-L. Definite-time relays control the reclosing of accelerating contactors, so that curve 1-L cannot be reached until after the motor has been retarded step by step through curves 4-L, 3-L, and 2-L, so that a high braking torque is available when the transfer to curve 1-L takes place.

In designing a d-c dynamic braking controller, care must be exercised in selecting the proper magnitude of d-c excitation current. This current must be sufficiently high to provide adequate maximum torque so that the motor will not run away with whatever maximum overload the hoist may be called upon to handle. A maximum braking torque of 1.7 times normal torque, which corresponds to approximately 2.5 times normal load on the hook, is usually ample for normal hook loads, with sufficient margin for handling reasonable overloads. On the other hand, the d-c excitation current should be kept low enough to prevent excessive heating of the motor. As an added refinement, d-c excitation may be controlled as a function of load, to strengthen the d-c flux with heavy loads.

SECONDARY-REACTOR CONTROL

Speed of a wound-rotor motor can be controlled by inserting a saturable reactor in the rotor circuit. The secondary-reactor method of wound-rotor motor control is illustrated in a schematic manner in Fig. 37-3. A saturable reactor is connected in series with the secondary resistor. The inductance of the reactor depends on its saturation. It is controlled by the amount of d-c excitation applied to its control winding. Increasing the d-c excitation decreases the inductance. Saturation, and consequently inductance, is also affected by the secondary current, that is, the motor torque. Increasing torque corresponds to decreasing inductance. The reactance of the reactor is likewise a function of motor speed. At low motor speeds, secondary frequency is high; and, for a given inductance, the reactance is high. With increasing motor speed, the secondary frequency decreases; hence the reactance, corresponding to a given inductance, decreases.

Adjusting the saturation of the secondary reactor by adjusting the excitation of its d-c control winding obtains speed control without secondary contactors. Space and maintenance are saved thereby. Since d-c excitation of the reactor is at a low power level, speed control can be accomplished with small control-circuit devices. A multiplicity of speed points can be provided, and practically stepless speed

SECONDARY-REACTOR CONTROL

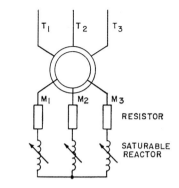

Fig. 37-3 Wound-rotor-motor speed control by saturable reactor in secondary circuit.

control is possible. Regulators may be introduced for adjusting and maintaining speed over a wide range of loads.

The general shape of speed-torque curves which can be obtained with secondary-reactor control is indicated in Fig. 37-3B. Curves are numbered 1, 2, 3, 4, corresponding to increasing amounts of d-c excitation of the reactor control winding. At high motor speeds, namely with low secondary frequency, secondary impedance is largely resistance and the effect of the reactor is small. The speed-torque curve is essentially a straight line. At lower speeds, namely with rising secondary frequency, the effect of the reactor becomes more pronounced and the speed-torque curves bend away from the straight line.

While Fig. 37-3 indicates the general shape of speed-torque curves obtained with secondary-reactor control, the exact shape of these curves will depend on the specific resistance and inductance of a motor

CONTROL OF WOUND-ROTOR MOTORS

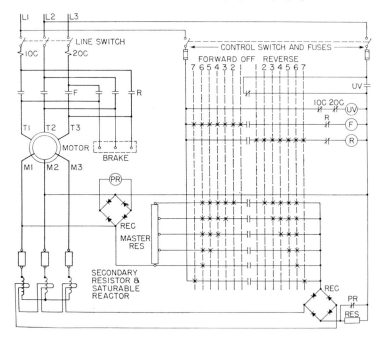

Fig. 38-3 Reversing plugging crane travel controller with secondary-reactor control. F = forward contactor; R = reverse contactor; UV = undervoltage relay; PR = plugging relay; $1OC, 2OC$ = overcurrent relays.

as well as on the reactor windings. Speed-torque characteristics must be determined for each type of motor by test or computation; it is not possible to establish general design data applicable to all kinds of wound-rotor motors. This means that the control designer must obtain from the motor designer recommendations on suitable resistance and inductance to meet specific application requirements.

The practical application of secondary-reactor control to a reversing plugging crane travel controller is illustrated in Fig. 38-3. Primary reversing control of the motor is obtained by reversing contactors F and R, which close when the master switch is turned in the forward or reverse direction. A semiconductor rectifier is used as a power source for the excitation of the saturable secondary reactor.

On the first point of the master switch in either direction, the d-c winding of the saturable reactor is open-circuited, and inductance of the reactor is maximum. As the master switch is advanced from point to point, the d-c winding of the reactor is connected to the rectifier,

SECONDARY-REACTOR CONTROL

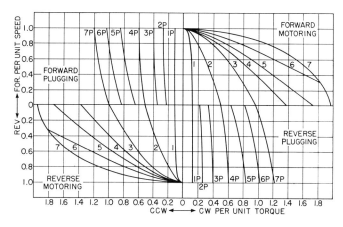

Fig. 39-3 Speed-torque curves of reversing plugging crane controller with secondary-reactor control.

and resistance in the d-c excitation circuit is short-circuited in steps. Thus the saturation of the reactor is gradually increased, resulting in a corresponding decrease in the inductance of the reactor. As a consequence, the impedance of the motor secondary circuit is reduced as the master switch is advanced.

The resulting speed-torque curves are illustrated in Fig. 39-3. The curves are symmetrical for the two directions of motor rotation. Quadrants 1 and 3 are the power driving quadrants in the two directions of rotation. Quadrants 2 and 4 are the plugging quadrants in which the drive is retarded by countertorque. The shape of the plugging curves is very desirable, since the variation in plugging torque over the speed range is moderate.

During plugging operation in quadrants 2 and 4, the saturation of the reactor is under the control of plugging relay PR. This relay is connected, through a semiconductor rectifier, to the secondary circuit of the motor; thus it responds to motor speed. As long as the motor is plugged, secondary voltage is higher than standstill voltage, PR is picked up, and a block of resistance is inserted in series with the rectifier supplying the reactor control winding. On each master-switch point, reactor saturation is reduced to a value below the saturation corresponding to operation in quadrants 1 and 3. When the motor reaches standstill, PR drops out and the plugging resistor is short-circuited.

A similar system of reactor secondary control has been developed

146 CONTROL OF WOUND-ROTOR MOTORS

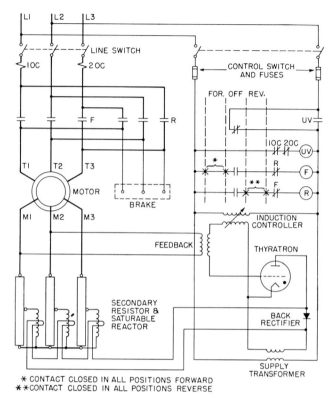

Fig. 40-3 Principal circuits of Harnischfeger Corporation reversing plugging crane controller with secondary-reactor control. F = forward contactor; R = reverse contactor; UV = undervoltage relay; $1OC$, $2OC$ = overcurrent relays.

by the Harnischfeger Corporation under the name of Electronic Stepless Control. The principal circuits of this type of control are illustrated in Fig. 40-3. Primary connections of the motor are reversed by reversing contactors F and R, which are closed when the master switch is turned in the forward or reverse direction. The saturable reactor is connected in parallel with a portion of the secondary resistor. Changing the saturation of the reactor changes the impedance of the motor secondary circuit. The d-c winding of the saturable reactor is excited from a thyratron and a half-wave semiconductor rectifier, which provide the required source of direct current. The voltage applied to the reactor d-c winding, hence the magnitude of the d-c excitation current, is varied by grid control of the thyratron.

PRIMARY-REACTOR CONTROL

Coupled to the master-switch handle is an induction controller which serves to vary the voltage applied to the thyratron grid. This induction controller is a transformer having a primary and secondary winding. A split rotating core, which is turned by the handle, varies the flux linkages between the primary and the secondary winding, hence the magnitude of voltage induced in the secondary winding. By turning the induction controller, it is possible to vary the grid voltage of the thyratron from zero to a maximum value, and correspondingly to vary the saturation of the reactor, thereby providing an infinite number of steps between minimum and maximum speed for any given load. For each position of the master switch, a fixed amount of reactor d-c excitation is obtained; and the motor performance is according to speed-torque curves which, in their general shape, resemble those in Fig. 39-3. Because of the stepless control afforded by the induction controller, an infinite number of speed-torque curves are available, and any operating point can be realized within an envelope similar to the one formed by curves 1 and 7 in Fig. 39-3 for each direction of rotation. A regulator, not shown in Fig. 40-3, responds to a motor secondary voltage feedback, that is motor speed, and introduces corrective voltage in the thyratron grid circuit which reduces the slope of the speed-torque curves, resulting in a lesser variation of speed with variations in torque. The secondary voltage feedback also is used to limit current peaks during acceleration and plugging, in case the induction controller is turned too fast. This feedback keeps the rate of change of motor speed within limits, thereby assuring that current and torque peaks stay within the capabilities of the motor and the drive.

PRIMARY-REACTOR CONTROL

Saturable reactors can be used to reverse motor rotation by changing the polarity of motor terminals with respect to the incoming line phases. It is also possible to vary the magnitude of voltage applied to the motor terminals. Hence it is possible not only to change direction of motor rotation, but also magnitude of motor speed and torque without the use of contactors in the power circuit. Figure 41-3 indicates in a schematic manner the primary power circuit of a primary-reactor reversing controller.

X-1 and X-2 are saturable reactors whose inductance is varied by control of the d-c excitation of the control winding. XT is a saturable transformer. It has a primary a-c winding XT-P and a secondary winding XT-S. The ratio of transformer primary turns to secondary

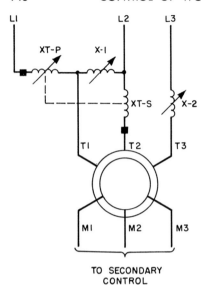

Fig. 41-3 Principal connections of motor reversing circuit with primary saturable reactors and transformer.

turns is 1 to 1. The two windings of the transformer are connected with polarities as indicated by the black squares. The voltage induced in XT-S is equal in magnitude but in opposite direction to the voltage impressed on XT-P. Secondary control of the motor consists of an impedance connected to the slip rings, either a resistor or a combination resistor and reactor.

How the direction and magnitude of voltages impressed upon the motor terminals are varied by controlling the saturation of the reactors and the transformer is illustrated in Fig. 42-3. The location of the various windings in the motor primary circuits is indicated at the top which corresponds to Fig. 41-3.

To obtain rotation in the forward direction, reactors X-1 and X-2 are saturated, whereas transformer XT is unsaturated. The drop across X-1 is small, and the voltage appearing across the transformer primary winding XT-P is nearly equal to the voltage between phases $L1$ and $L2$. Thus motor terminal $T1$ is nearly at the same potential as line terminal $L2$. The voltage appearing across the transformer secondary XT-S is of the same magnitude as the voltage across the transformer primary winding, but opposite in direction. Hence motor terminal $T2$ is at nearly the same potential as line terminal $L1$. Since the drop across reactor X-2 is small, motor terminal T-3 is at nearly the same potential as line terminal L-3. Thus the voltage triangle im-

PRIMARY-REACTOR CONTROL 149

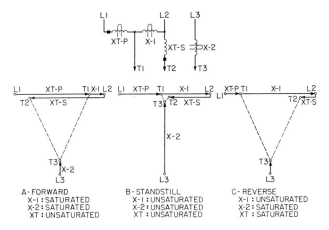

Fig. 42-3 Drops and motor voltages occurring during reversing with primary saturable reactors and transformer.

pressed upon the motor primary winding is substantially equal to the line-voltage triangle.

A gradual reduction of current in the d-c windings of reactors X-1 and X-2 increases their reactance, and hence the drop across their a-c windings. Figures 42-3B shows that, with reactors and transformer unsaturated, the voltage triangle impressed upon the motor terminals has shrunk to a very small magnitude. The motor develops practically no torque; and with no overhauling torque acting on the shaft, the motor is at standstill. The motor acts as if its line contactor had been opened and disconnected it from the line.

What happens when reactor X-2 and transformer XT are saturated, while reactor X-1 is unsaturated, is indicated in Fig. 42-3C. The drop across reactor X-1 is large, whereas only a small voltage appears across the primary winding of the transformer XT-P. This means that motor terminal $T1$ is substantially at the same potential as line terminal $L1$. Since the voltage induced in the transformer secondary winding XT-S is also small, the potential appearing on motor terminal $T2$ is substantially the same as that of line terminal $L2$. Because of the small inductance of reactor X-2, motor terminal T-3 is substantially at the same potential as line terminal $L3$. The voltage triangle appearing across the motor primary winding is substantially the same as indicated in Fig. 42-3A, except that the polarity of terminals $T1$ and $T2$ has been interchanged. That is to say, the motor is now operating in the reverse direction of rotation.

Varying the current in the d-c control windings of the reactors and the transformer permits adjustment of the voltages impressed upon the motor terminals with respect to magnitude and phase rotation. Since a small amount of d-c control power suffices to control the saturable reactors and transformer, it is possible to use small control devices to adjust the saturation in a practically infinite number of steps. One method is to excite the d-c control windings from a constant-voltage d-c source and to use a rheostat with a large number of steps for adjusting the current in the d-c control windings. Another method is to use an induction controller that produces an adjustable a-c control voltage with an infinite number of steps, which is then rectified and impressed on the d-c control windings. It is also possible to introduce regulators which regulate any desired operating quantity, such as maintaining constant speed over a wide range of torques.

REACTOR CRANE HOIST CONTROL

The use of reactors for the primary and secondary control obtains motor performance which is well suited for material-handling equipment, particularly for crane hoist drives. With this system of control, it is possible to obtain satisfactory performance in all four quadrants of motor performance in which the motor may be called upon to drive a load or to restrain an overhauling load. Using a speed regulator for the control of the saturable reactor and transformer d-c windings, it is possible to obtain substantially constant-speed operation over a wide range of load torques. In that respect, reactor control of wound-rotor motors obtains performance characteristics which approach closely those which can be obtained with adjustable-voltage control of d-c motors.

The power circuit of a typical reactor crane hoist controller is indicated in Fig. 43-3. The primary motor circuit contains two saturable reactors and a saturable transformer. These components serve to reverse the motor primary and to modulate the magnitude of the voltage applied to the motor stator terminals. A line switch isolates the motor from the line. A line contactor, included as a safety measure, serves to disconnect the motor from the line in case of overcurrent or loss of voltage. It provides undervoltage protection and can also be utilized for automatic stopping of the load under the control of limit switches. The line contactor is not necessary for stopping a load; the operator may "float" a load by manipulating the master switch without removing power from the motor or setting the brake. The secondary circuit of the motor contains a network consisting of a fixed

REACTOR CRANE HOIST CONTROL

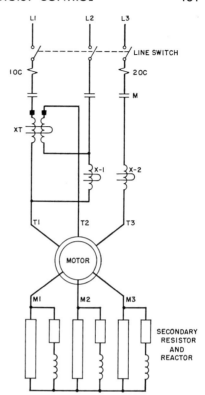

Fig. 43-3 Power circuits of primary and secondary reactor crane hoist controller. M = line contactor; $1OC$, $2OC$ = overcurrent relays; X-1, X-2 = saturable reactors; XT = saturable transformer.

resistor and reactor in each phase. This secondary impedance network molds the motor speed-torque curves so that satisfactory acceleration from standstill to full speed can be obtained without the use of secondary contactors.

The kind of motor speed-torque curves obtained with the reactor control for one direction of voltage phase sequence is illustrated in Fig. 44-3. The speed-torque curve obtained with full voltage applied to the motor stator terminals has the same general shape as the curves previously discussed in connection with Fig. 39-3. As the motor terminal voltage is reduced by changing the d-c excitation of the reactors and transformer, a family of curves is obtained which are similar in shape but with reduced torque corresponding to each value of speed. The most desirable performance of a crane hoist controller is to obtain on each master-switch point a definite speed which is substantially independent of the load on the crane hook. From Fig. 44-3 it can be seen that, for any amount of torque on the motor shaft, there is a

152 CONTROL OF WOUND-ROTOR MOTORS

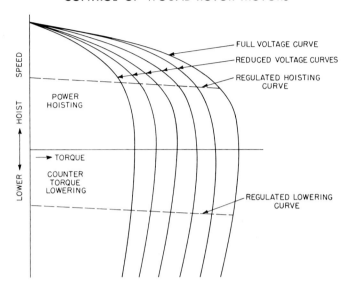

Fig. 44-3 Performance of wound-rotor hoist motor with primary- and secondary-reactor control.

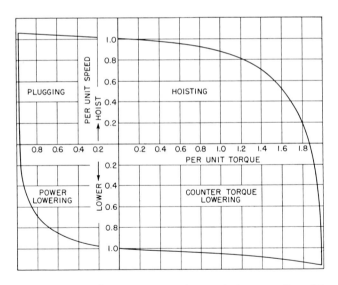

Fig. 45-3 Envelope of performance range of crane hoist controller with primary- and secondary-reactor control.

PRIMARY-REACTOR AND VOLTAGE-UNBALANCE CONTROL 153

speed-torque curve, that is to say, a certain condition of current in the d-c control windings of saturable reactors and transformer which corresponds to a desired motor speed, either hoisting or lowering. This performance characteristic can be realized by using a speed regulator, such as a magnetic amplifier, which adjusts the d-c current in the reactor and transformer control windings so as to obtain constant motor speed over a torque range within the capabilities of the motor.

One practical method of speed control is to use a master switch coupled to a potentiometer rheostat. When the master-switch handle is turned, the rheostat establishes a d-c reference voltage in a positive or negative direction, which is fed into the regulator as a command representing the desired hoisting or lowering speed. A tachometer generator, coupled to the motor shaft or to one of the drive shafts, provides the feedback to the regulator, indicating actual motor speed. The regulator then produces, through a suitable amplifier, a d-c output which obtains correct d-c excitation of the reactors and transformer in the motor primary circuit, so that actual motor speed corresponds to the command established by the master-switch position.

This type of control obtains practically stepless adjustment of motor speed within the range of speed and torque capabilities of the motor. The number of steps is equal to the number of contact points on the control rheostat. Figure 45-3 indicates the envelope of the area of possible performance which can be covered with a primary- and secondary-reactor crane hoist controller. Normal loads are hoisted by applying power to the load in quadrant 1. Plugging torque in quadrant 2 is available for stopping an empty hook or a light load rapidly. Overhauling loads are lowered by countertorque in quadrant 4. Power lowering torque is available in quadrant 3 for accelerating an empty hook or a light load in the lowering direction.

PRIMARY-REACTOR AND VOLTAGE-UNBALANCE CONTROL

A combination of primary-reactor reversing and speed control by primary-voltage unbalancing is utilized by the Westinghouse Electric Corporation for their Load-O-Matic * line of crane control. The power circuit of a Load-O-Matic crane hoist controller is shown in Fig. 46-3. A line switch is provided for isolation, and a line contactor is included as a protective measure against overload, undervoltage, and overtravel when overtravel limit switches are used. The line contactor is not needed for spotting the load and stopping the motor. Motor

* Registered trade mark of the Westinghouse Electric Corporation.

154 CONTROL OF WOUND-ROTOR MOTORS

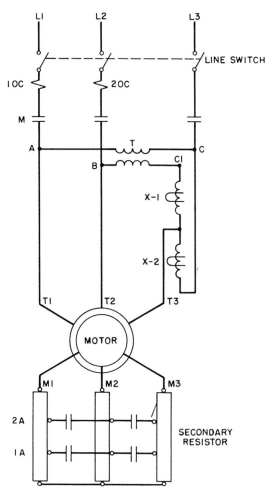

Fig. 46-3 Power circuits of Westinghouse Electric Corporation Load-O-Matic crane hoist controller. M = line contactor; $1A$, $2A$ = secondary accelerating contactors; $1OC$, $2OC$ = overcurrent relays; X-1, X-2 = saturable reactors; T = transformer.

primary terminals $T1$ and $T2$ are always connected to line phases $L1$ and $L2$. Motor primary terminal $T3$ is connected to a network which includes a transformer T having a 1-to-1 turn ratio of primary and secondary windings, and saturable reactors X-1 and X-2. This network serves to shift the potential on terminal $T3$, thus determining phase rotation and degree of unbalance of the primary voltages ap-

PRIMARY-REACTOR AND VOLTAGE-UNBALANCE CONTROL 155

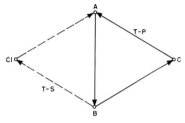

A—TRANSFORMER T ESTABLISHES POINT CI

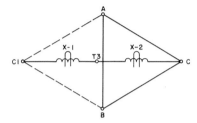

Fig. 47-3 Effect of transformer and saturable reactors of Load-O-Matic controller on unbalancing motor primary voltage.

B—REACTORS X-I AND X-2 DIVIDE VOLTAGE C-CI

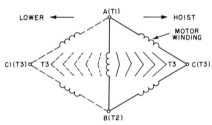

C—UNBALANCING MOTOR VOLTAGE BY SHIFTING POTENTIAL OF T3

plied to the motor windings. The secondary circuit of the motor contains a resistor and two accelerating contactors.

The connection of the transformer and the saturable reactors is illustrated in Fig. 47-3. The solid right-hand triangle in sketch A represents the voltages existing at points A, B, C (see Fig. 46-3), which are at the same potentials as line terminals $L1$, $L2$, and $L3$. This triangle represents a balanced positive-sequence voltage system obtaining motor rotation in the hoisting direction. The primary winding of the transformer T-P is connected between points A and C; thus full line-to-line voltage is applied to the transformer. A voltage of equal magnitude appears across the transformer secondary winding T-S. By connecting the transformer secondary winding to point B and selecting the proper phase relationship between primary and secondary

CONTROL OF WOUND-ROTOR MOTORS

transformer windings, a voltage triangle A, B, $C1$ is created which is represented by the broken lines in Fig. 47-3A. This triangle is of the same magnitude as the line-voltage triangle, but it has negative phase sequence. If this triangle is applied to the motor primary, motor rotation in the opposite direction, that is in the lowering direction, results.

The two saturable reactors X-1 and X-2 are connected between points $C1$ and C. Motor terminal $T3$ is connected to the midpoint between the two reactors. As can be seen from Fig. 47-3B, the saturable reactors act as a voltage divider between points $C1$ and C. By varying the saturation, hence the reactance of the reactors, motor terminal $T3$ can be shifted between points $C1$ and C.

The voltage triangle impressed upon the motor primary terminals for various degrees of saturation of the reactors is indicated in Fig. 47-3C. Terminals $T1$ and $T2$ are always connected to points A and B, and full line voltage appears across these terminals all the time. With reactor X-2 fully saturated, the drop across that reactor is small, and motor terminal $T3$ is practically at the same potential as point C. Full positive-sequence voltage is applied to the motor, which acts like a balanced motor turning in the hoisting direction. Reducing the saturation of reactor X-2 causes the drop across this reactor to increase, and the potential of terminal $T3$ moves to the left, that is toward $C1$. The voltage triangle, applied to the motor, becomes progressively unbalanced, and the amount of negative-sequence voltage increases with decreased saturation of X-2. However, the positive-sequence voltage remains larger than the negative-sequence voltage. With both reactors unsaturated, the potential of terminal $T3$ falls midway between points A and B. Positive- and negative-sequence voltages are equal. The motor acts like a single-phase motor and develops no torque at standstill. Saturating reactor X-1 causes the potential of terminal $T3$ to move further toward $C1$. Negative-sequence voltage increases and overcomes positive-sequence voltage. With X-1 fully saturated, the potential of terminal $T3$ becomes practically the same as $C1$, which means that the motor acts like a balanced motor with reversed phase sequence.

Semiconductor rectifiers supply the d-c power required for the control windings of the saturable reactors. The master switch having four positions in the hoisting and lowering direction is used to switch resistors in the reactor control windings and to actuate the secondary accelerating contactors. Typical speed-torque curves obtained with a Load-O-Matic crane hoist controller are shown in Fig. 48-3. In

PRIMARY-REACTOR AND VOLTAGE-UNBALANCE CONTROL 157

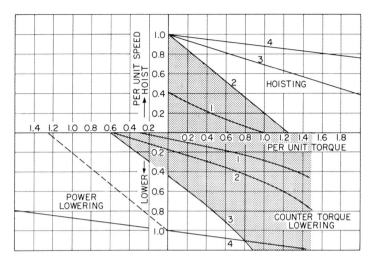

Fig. 48-3 Speed-torque curves of Westinghouse Electric Corporation Load-O-Matic crane hoist controller.

the hoisting direction, point 1 of the master switch applies unbalanced voltage to the primary with all resistance in the secondary circuit. The resultant speed-torque curve obtains a reduced no-load speed. On point 2, practically balanced positive-sequence voltage is applied to the motor. On points 3 and 4, secondary contactors $1A$ and $2A$ close and accelerate the motor to full speed. In the lowering direction, curves 1, 2, and 3 are obtained with unbalanced voltage. A tachometer generator connected to the hoist motor supplies a corrective voltage to the reactor control winding to obtain some flattening of the curves. On the fourth point, full negative-sequence voltage is applied to the motor, contactor $2A$ closes, and the load is lowered by regenerative braking.

A vernier attachment is available by which the operator may obtain fine speed adjustments between master-switch positions. The vernier is a rotary transformer, independent of the master switch, which is operated by hand or foot. It feeds a control signal into the saturable reactors. If the operator desires to spot the load, he may leave the master switch in any of the first two positions hoisting or the first three positions lowering, and then spot the load accurately with the vernier. The area in which stepless control can be obtained by the vernier is indicated in Fig. 48-3 by shading.

SPEED CONTROL BY SECONDARY POWER CONVERSION

The speed-control systems described so far utilize a fixed or variable external impedance in the rotor circuit, the drop across which determines the secondary voltage appearing across the motor slip rings—hence the motor speed. Motor speed can also be controlled by introducing into the rotor circuit, from an external source, a voltage of proper magnitude, phase angle, and frequency. A separate power source, generally additional rotating machinery, is required to produce the external secondary voltage. Since power flowing out of or into the slip rings of the induction motor has a voltage and frequency depending on the design characteristics of the motor and its speed, the external power apparatus must be capable of converting the slip power of the induction motor into some other suitable d-c or a-c power. Induction motor speeds below synchronism are obtained by introducing into the secondary circuit a voltage which adds to the voltage drop in the rotor windings. Power flow is from the induction motor to the power conversion apparatus. Speeds above synchronism are obtained by impressing on the induction motor rotor a voltage in a sense opposite to the rotor voltage drop. Power flow is then from the conversion apparatus to the motor rotor.

By varying not only the magnitude but also the phase angle of the voltage impressed on the rotor, the phase angle of the secondary current can be shifted so that a portion of the magnetizing current of the motor is supplied from the rotor circuit instead of the stator. This shift results in an improvement of the power factor of the motor, as viewed from the power system.

The size of the power-conversion set is determined by the slip power it must handle. This power is equal to the rating of the induction motor, multiplied by the maximum slip for which the equipment is designed. Because of the cost of the additional machinery, rotating power-conversion sets for speed control are generally economical only for large motors and for a limited speed range. If a speed range above 2 to 1 is required, or if motors of small and medium size are involved, the use of a-c to d-c conversion equipment and adjustable-voltage d-c drives is generally more economical.

Secondary power-conversion sets for use with large induction motors enjoyed considerable popularity during the 1920's for large-horsepower drives, such as those used in steel-rolling mills. Today, they have been largely superseded by adjustable-voltage d-c drives, partly because of increased economies in designing and building d-c machinery, partly

KRAEMER SYSTEMS

because of the more exacting present-day demands on speed control, which can be met more easily with adjustable-voltage d-c drives and high-performance speed regulators. However, even today large induction motors with secondary power-conversion sets offer economic advantages on drives with fan-load characteristics, such as wind-tunnel drives, dredge pumps, and the like, because for such loads the secondary power-conversion sets are smaller than for constant-torque loads. In the following paragraphs a number of secondary power-conversion systems are described which have found practical use.

KRAEMER SYSTEMS

Kraemer systems use d-c machines to convert slip power and to supply the voltage impressed on the rotor of the main induction motor. Shown in Fig. 49-3, are the basic circuits of two systems. The slip rings of the main motor are connected to the slip rings of a rotary converter, the commutator of which is connected to the commutator of a d-c motor. In sketch A, the d-c motor is coupled to the main motor shaft. Thus the slip power from the main motor rotor flows through the rotary converter into the d-c motor and is utilized as shaft output of the main motor. This equipment is capable of producing constant

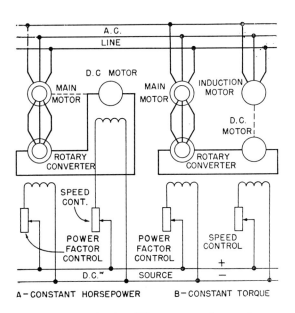

Fig. 49-3 Principal circuits of Kraemer speed-control systems.

160 CONTROL OF WOUND-ROTOR MOTORS

horsepower over its speed range. In sketch B, the d-c motor is coupled to an induction motor, which is driven as an induction generator and feeds the slip power back into the a-c line. This equipment is able to operate over its speed range at constant torque. A constant-voltage d-c supply must be available to furnish excitation for the synchronous converter and the d-c motor.

The speed of the main motor is varied by changing the excitation of the d-c motor, which in turn controls the voltage appearing across the commutator of the rotary converter. This voltage determines the a-c voltage across the rotary converter slip rings, which is impressed on the main motor slip rings. The speed range which can be obtained with the Kraemer system is limited only by the cost involved, as with sufficiently large d-c machines it would be possible to reduce the main motor speed to zero. Power-factor correction can be obtained by overexciting the rotary converter. Design limitations of the rotary converter make it impractical to improve the main motor power factor beyond approximately 0.95.

To start a Kraemer system, the main motor is first started and brought to normal speed with a conventional resistance-type starter. In a constant-torque system, the d-c induction motor set is also started. With no excitation on the d-c motor, the slip rings of the main motor are transferred from the starting resistor to the rotary converter. Excitation is then applied to the d-c motor, and the main motor assumes a speed corresponding to the voltage produced by the d-c motor.

Because of the low slip power available near synchronous speed of the main motor, the rotary converter becomes unstable at low slip frequency. This limits Kraemer system operation to speeds obtaining frequencies above two cycles, so that the Kraemer system is used only for subsynchronous operation. By reversing the excitation of the d-c motor, it is possible to accelerate an unloaded Kraemer set through synchronous speed and to obtain stable operation above synchronous speed without difficulty. However, the "dead zone" of two cycles below and above synchronous speed, within which the main motor cannot deliver any appreciable amount of torque, makes operation of Kraemer systems above synchronism impracticable.

Recent advances in semiconductor rectifiers have made their use attractive as power-conversion units in a modified Kraemer system which is built by Westinghouse Electric Corporation under the name of Rectiflow * drive. Principal circuits of this system are indicated in

---

* Registered trade mark of the Westinghouse Electric Corporation.

KRAEMER SYSTEMS

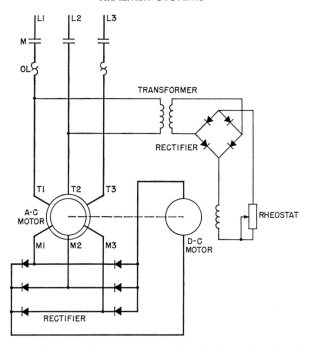

Fig. 50-3 Principal circuits of Westinghouse Electric Corp. "Rectiflow" drive with secondary power conversion. M = line contactor; OL = overload relays.

Fig. 50-3. A d-c motor is coupled to the shaft of the a-c motor. The slip rings of the a-c motor are connected to a three-phase full-wave rectifier, which converts the slip power into direct current and feeds it into the armature of the d-c motor. Another single-phase, full-wave rectifier supplies excitation for the d-c motor field. Speed of the a-c motor is controlled by the field rheostat of the d-c motor. The rectifier and armature of the d-c motor act like a small amount of permanent starting resistance in the a-c motor secondary curcuit. The starting current of the a-c motor is comparable to the starting current of a squirrel-cage motor of the same horsepower rating. Small motors can be started with a conventional full-voltage starter. Larger motors can be started by part-winding starting or any of the conventional reduced-voltage starters. Rectiflow drives are available in ratings as low as 7.5 horsepower.

CONTROL OF WOUND-ROTOR MOTORS

SCHERBIUS SYSTEMS

In a Scherbius system, the slip rings of the main motor are connected to the commutator of a so-called "Scherbius machine." This machine is built similar to a conventional d-c machine. Three sets of brushes are arranged on the commutator so that three-phase power can be taken off. The field structure is excited at slip frequency through an autotransformer. Thus the frequency of the voltage appearing at the commutator brushes equals slip frequency. By changing taps of the autotransformer, the magnitude of the voltage across the commutator of the Scherbius machine, and correspondingly the slip of the main motor, can be varied. By suitable design of the Scherbius machine, the phase angle of its commutator voltage can be deflected so as to improve the power factor of the main motor to approximately 0.95.

Figure 51-3A shows the basic connections of a constant-torque Scherbius set. The Scherbius machine is coupled to an induction machine. Slip power is transmitted from the main-motor rotor through the Scherbius machine to the induction machine, which feeds the energy back into the line. Therefore, this system operates at constant torque over its speed range. It would be possible to couple the Scherbius machine directly to the shaft of the main motor. Slip power would then be transmitted back to the main-motor shaft as mechanical power, and the system would operate over its speed range at constant horsepower.

Because of commutation limitations of Scherbius machines, the possible reduction in speed below synchronous speed is limited to a slip frequency of 18 cycles. When synchronous speed is approached, slip frequency and slip power decrease. At synchronism, slip frequency and voltage across the Scherbius machine are zero, and no voltage is available to overcome the drop across the internal resistance of the main motor rotor. The system according to Fig. 51-3A cannot operate at, nor be accelerated through, synchronous speed. This system can operate only below synchronous speed. It is a "single-range" system.

It is possible to modify a Scherbius system so that it can accelerate through synchronous speed and operate below and above synchronous speed as a "double-range" system. To accomplish this operation, an additional machine is used, called the "ohmic drop exciter." The basic circuits of this machine, which is coupled to the main motor, are indicated in Fig. 51-3B.

The ohmic drop exciter consists of an armature with slip rings at one end and a commutator at the other end. There is no stator. The slip rings are excited at line frequency through a transformer. The

SCHERBIUS SYSTEMS 163

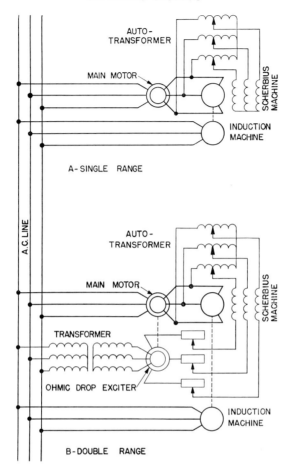

Fig. 51-3 Principal circuits of Scherbius speed-control systems.

ohmic drop exciter acts as a frequency changer, and a voltage at slip frequency appears across the commutator. This voltage is fed to the field winding of the Scherbius machine. A resistor in series with the ohmic drop exciter is adjusted so that the exciter voltage equals the drop across the resistor and the Scherbius machine field winding. At synchronous speed of the main motor, the ohmic drop exciter produces a d-c voltage sufficient to excite the Scherbius machine, which in turn generates sufficient voltage to overcome the internal resistance drop in the main-motor rotor winding. Stable operation at synchronous speed is possible.

As compared with a Kraemer system, the size of the power-conversion machines in a Scherbius system is smaller for a given total speed range, since the range of the Scherbius set can be divided into half above and half below synchronous speed. That is, for a given total speed range, the Scherbius system operates only at half slip, either plus or minus.

FREQUENCY-CHANGER SYSTEMS

Basic circuits of speed control systems with frequency changers are indicated in Fig. 52-3. Diagram A illustrates a constant-torque system. The slip rings of the main motor are connected to the commutator of a frequency changer, the slip rings of which are connected to the line through a tap-changing transformer. The frequency changer is built like an ohmic drop exciter, except that it has a larger rating. Coupled to the main motor is a small synchronous generator which produces a frequency corresponding to actual motor speed. To this generator is connected a small synchronous motor which drives the frequency changer and supplies its losses. Thus the frequency changer rotates at a speed corresponding to actual motor speed, and the frequency of the voltage across the frequency-changer slip rings is equal to line frequency.

Slip power flows from the main-motor slip rings through the frequency changer and back into the line. Hence this set operates at constant torque over its speed range. Changing taps on the transformer varies the voltage across the frequency-changer slip rings. This voltage in turn affects the voltage across the frequency-changer commutator and the speed of the main motor. Some degree of power-factor correction can be obtained by shifting the commutator brushes of the frequency changer.

At synchronous speed of the main motor, the frequency changer is driven at synchronous speed and direct current appears across its commutator, overcoming the resistance drop in the main-motor rotor. Stable operation at synchronous speed is possible, and frequency-changer systems can be built for double-range operation. The speed range obtained is determined by the commutating limit of the frequency changer.

Frequency-changer systems can also be built for constant-horsepower operation, as indicated in Fig. 52-3B. The slip rings of the frequency changer are connected to a synchronous motor, coupled to the shaft of the main motor. The driving motor of the frequency changer is connected to the line, and the frequency changer runs at

DOUBLY FED INDUCTION MOTOR

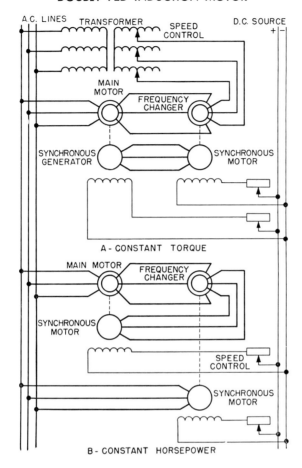

Fig. 52-3 Principal circuits of frequency-changer speed-control systems.

synchronous speed. Slip power is fed from the frequency-changer slip rings to the synchronous motor and is utilized as mechanical power through the main-motor shaft. The voltage across the frequency changer and, correspondingly, the speed of the main motor are varied by adjusting the field excitation of the synchronous motor.

DOUBLY FED INDUCTION MOTOR

Speed adjustment of a wound-rotor motor can be obtained by connecting the stator to a line of constant frequency and the slip rings to a synchronous generator supplying adjustable frequency. Such an ar-

CONTROL OF WOUND-ROTOR MOTORS

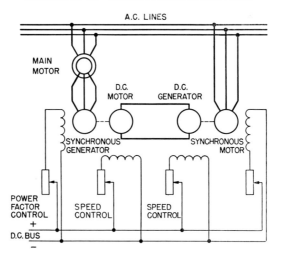

Fig. 53-3 Doubly fed wound-rotor induction motor with secondary power-conversion set.

rangement is shown in Fig. 53-3. The main-motor slip rings are connected to a synchronous generator, which is driven by an adjustable-speed d-c motor. This motor may be a conventional d-c shunt motor, provided a suitable supply of d-c power is available; or it may be connected to an adjustable-voltage d-c generator, which in turn is driven by a synchronous or induction motor. Varying the speed of the d-c motor by adjusting the d-c generator voltage varies the frequency of the synchronous generator, and consequently the main-motor speed. The power factor of the main motor can be varied by adjusting the excitation of the synchronous generator. Slip power flows from the main motor through the synchronous generator and the adjustable-voltage motor-generator set back to the line. This is therefore essentially a constant-torque system.

As compared with other speed-control systems, a doubly fed induction motor system employs more rotating machinery. However, several operating advantages offset this disadvantage. Since no commutator-type machines are used in the main-motor secondary circuit, the speed range is not limited. It is possible to reduce the main-motor speed to standstill. Therefore, this system can be used for the starting of very large motors, having ratings of the order of tens of thousand horsepower, with minimum line disturbance. With the main-motor

POWER SELSYN OPERATION 167

stator disconnected from the line, the adjustable-voltage d-c drive is started, and the frequency of the synchronous generator is brought to line frequency. With the main motor at standstill, line frequency then appears across the stator terminals, and the main-motor stator can be synchronized with the line. After the stator has been connected to the line, the main motor is accelerated by reducing the speed of the synchronous generator.

As the speed of the main motor is determined by the d-c adjustable-voltage set, it is easy to arrange for accurate automatic speed control. All the refinements that can be built into adjustable-voltage control systems can be applied to this type of induction-motor speed-control system. It is also possible to regulate the main-motor power factor by automatic control of the field excitation of the synchronous generator.

POWER SELSYN OPERATION

When two wound-rotor induction motors are interconnected so that corresponding phases of their stator windings are connected to the same phases of the line, and corresponding rotor phases are interconnected, the rotor voltages of both motors must be of equal magnitude, and both motors must operate at the same per unit speed. As long as the two rotor voltages have the same phase angle, the two rotors will retain the same angular position in space. This means that the two motors will run in synchronism. If one of the two rotors lags in angular position with respect to the other, a phase shift occurs between the two rotor voltages which, in turn, causes a circulating current to flow between the two rotors. This current develops a torque in the direction to reduce the angle between the two rotors. This so-called "synchronizing" torque keeps the two rotors in step as long as the maximum torque which the combination is able to develop is not exceeded. Such a system of induction motors is described as a "power selsyn" (from *self-syn*chronizing) system. Some authors use the name "synchrotie" system because two machines are tied together in synchronism.

Connections of a simple power selsyn system are shown schematically in Fig. 54-3. It consists of two wound-rotor motors which may or may not have an equal number of poles. Their rotor open-circuit voltages must be the same, so that the rotor voltages match at any speed. If one of the induction motors, the transmitter, is driven by a motor, it transmits torque to the second induction motor, the receiver, which rotates at the same relative speed (as determined by its number of poles) as the transmitter. Power selsyns can be used to operate two

CONTROL OF WOUND-ROTOR MOTORS

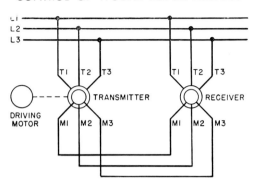

Fig. 54-3 Power selsyn system consisting of two wound-rotor induction motors.

mechanically separated drives at the same speed and in synchronism with each other by means of electrical interconnections only.

Power selsyns may be driven either in the same direction or in the opposite direction as the rotation of flux. The maximum torque that can be developed by the combination varies with relative speed. In Fig. 55-3 the torque of transmitter and receiver are plotted as a function of mechanical rotation with or against flux rotation. Base torque is the rated torque which the selsyn machines would develop when operated as induction motors. When selsyns are operated in the same direction as the flux, the synchronizing torque decreases rapidly as synchronous speed is approached. For this reason, selsyns turning in the same direction as flux rotation should not be operated at speeds above two thirds synchronous speed. Selsyns turning against flux rotation can be operated at higher speed, even above synchronism. The top speed is limited by the induced rotor voltage, by mechanical

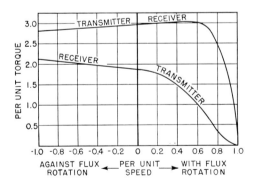

Fig. 55-3 Maximum-torque characteristic of two identical wound-rotor motors operating as power selsyns.

POWER SELSYN OPERATION

strength of the rotors, and by heating resulting from rotor core losses. If a selsyn system is subjected to rapid changes in speed or in torque, the system may have a tendency to hunt. Whether or not the system will actually hunt depends largely on design constants of the selsyn machines and on the mechanical constants of the drive. If a system is found to hunt, this situation can be remedied by inserting resistance in the rotor or stator phases of the receiver. The maximum-torque capability of the receiver is reduced thereby, but the receiver follows speed changes of the transmitter with less tendency to overshoot and hunt.

Power selsyns may be used to hold two independent drive motors in step. Referring to Fig. 54-3, both the transmitter and the receiver may be connected to driving motors. The selsyn system, then, does not transmit all the required driving torque to the receiver, but only sufficient torque to keep the two drive motors in synchronism.

The synchronizing torque holding the selsyn motors in step varies with the angle of displacement between the two rotors. The angle of displacement is measured in electrical degrees: 360 electrical degrees are one revolution divided by the number of pairs of poles. Synchronizing torque is zero at zero-degree and approximately 180-degree displacement. The torque reaches maxima at approximately 90- and 270-degree displacement.

Before a power selsyn system can be started, it is necessary to synchronize or "lock in step" the selsyn motors at standstill. If three-phase voltage were applied and the selsyn motors were standing still with a large angular displacement between their rotors, a high torque would be developed, pulling the motors in step. However, in most instances, the motors would overshoot the locked-in-step rotor position and start to run as induction motors, failing to synchronize. If single-phase voltage is first applied, a synchronizing torque is developed which locks the rotors in step. The motors, however, cannot start with single-phase excitation, since the starting torque is zero. Power selsyn drives may therefore be started in such a manner that at first two phases of the transmitter and receiver are connected to the line so that in fact single-phase excitation is applied to both machines. Depending on the angular displacement between the two rotors, they may lock in step by turning either toward the zero-degree displacement position or toward the 180-degree displacement position. After a slight time interval, to allow for mechanical movement of the drive, the third phases of the transmitter and receiver are connected to the line. Three-phase excitation is then applied, and the two machines will operate in synchronism.

CONTROL OF WOUND-ROTOR MOTORS

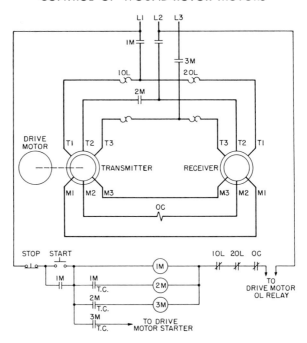

Fig. 56-3 Three-step starting control of power selsyn system. $1M, 2M, 3M$ = line contactors; $1OL, 2OL$ = overload relays; OC = overcurrent relay.

The above method, called the "two-step method," is quite satisfactory if the mechanical drive is such that the two selsyns are free to turn in either direction when seeking to lock in step. On some drives, the mechanical arrangement gives only freedom to move in one direction in order to lock in step. An example is a lift bridge, the two ends of which are driven by independent motors, tied together in step by power selsyns. When the bridge is seated and the power selsyns are to be locked in step prior to the movement of the bridge, the power selsyn rotors can turn only in one direction, namely so as to lift one end of the bridge off its seat. In such cases, it is preferable to employ the "three-step method." Connections for a simple three-step starter are given in Fig. 56-3, while Fig. 57-3 illustrates the synchronizing torques obtained on each of the three steps.

The diagram in Fig. 56-3 assumes that there is a drive motor, a transmitter, and a receiver. Pressing the start button energizes contactor $1M$, which applies single-phase voltage to the stators of both

POWER SELSYN OPERATION 171

selsyns, the third stator phase of each motor being left disconnected. The synchronizing torque reaches a maximum near 180-degree displacement, and the selsyn motors are pulled together in a direction toward the zero-displacement position. A time-closing interlock on $1M$ energizes contactor $2M$. On the second step, the third stator phases of the two selsyns are connected together. The torque curve approximates a sine wave with maxima at 90- and 270-degree displacement. The maximum torque, however, barely equals rated motor torque. After a timing interlock on $2M$ has closed, contactor $3M$ is energized. On the third step, the third stator phases are connected to the line, so that three-phase excitation is applied to the two selsyn motors, and the maximum torque rises considerably above the single-phase value. This increased synchronizing torque pulls the rotors together to less than a 30-degree displacement, the exact angle depending on the load

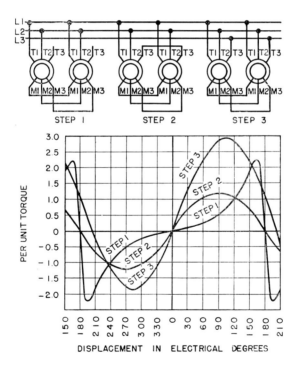

Fig. 57-3 Typical torque-angle curves for three-step starting sequence of power selsyn system.

172 CONTROL OF WOUND-ROTOR MOTORS

torque the selsyns have to overcome. The two selsyn motors are then firmly locked in step, and the selsyn system is ready to be accelerated. A time-closing interlock on $3M$ now permits the starting of the drive motor, either automatically or by some other pushbutton or master switch.

If the drive motor is started by a second manual command by the operator, care must be taken to prevent leaving the power selsyns energized at standstill for a prolonged period of time. To prevent overheating, a timing relay should be provided which deenergizes the selsyns if the drive motor is not started within a certain period of time. The connections, as shown in Fig. 56-3, imply that the selsyns are deenergized simultaneously with the drive motor. This means that when the drive motor is stopped, the selsyns will coast to a stop independently. On some applications, it may be desirable to maintain the selsyn tie while the drive is decelerating to a stop. The control can then be arranged so that, when the drive motor stops, the selsyns remain energized for a few seconds in order that they may come to a stop in synchronism.

Thermal overload relays are included to protect the transmitter and receiver individually against overloading. It is also advisable to provide an overcurrent relay in one of the rotor connections to protect the system in case the selsyns should pull out of step as a result of mechanical overload. Pulling out of step is accompanied by a high current surge. If the selsyns operate in the same direction as flux rotation, a pull-out would cause the receiver to speed up to its synchronous speed. If the selsyns are operating against flux rotation, a pull-out may cause the receiver to reverse. Either of these conditions can result in injury to personnel or damage to the machinery or the product.

Although a number of experimental circuits have been worked out to synchronize power selsyn motors while they are running, none of these is sufficiently simple and reliable to have found widespread application. Controllers must be so designed that the motors are synchronized at standstill, before the system starts to rotate.

Whereas the foregoing explanation of power selsyn operation is based on a system involving two selsyn motors, it is possible to build systems consisting of three or more selsyn motors. Either a multiplicity of individual motorized drives may be kept in synchronism by power selsyns, or one transmitter may transmit power to a multiplicity of receiver units. In a multiple-motor system, resistance should be inserted in the stator or rotor phases of the receiver units to reduce any tendency of the system to hunt because of unequal torque peaks to which the various receiver units may be subjected.

BIBLIOGRAPHY

1. D. R. Shoults, C. J. Rife, and T. C. Johnson (Editor), *Electric Motors in Industry,* Chapter 5, John Wiley and Sons, New York, 1942.
2. P. B. Harwood, *Control of Industrial Motors,* Chapter 17, John Wiley and Sons, New York, 1952.
3. R. W. Jones, *Electric Control Systems,* Chapters 4, 11, 12, John Wiley and Sons, New York, 1953.
4. L. A. Umansky, "Adjustable Speed Drives for Rolling Mills," *Iron and Steel Engineer,* September 1924.
5. J. A. Jackson, "Control for A.C. Grab-Bucket Hoists," *General Electric Review,* September 1926.
6. L. M. Nowacki, "Induction Motors as Selsyn Drives," *Electrical Engineering,* December 1933.
7. H. L. Wilcox, "Fast Crane Control," *Steel,* April 24, 1939.
8. D. C. Wright, "Frequency Relays and Motor Acceleration for Steel Mill Control," *Iron and Steel Engineer,* May 1939.
9. C. C. Clymer, "Large Adjustable-Speed Wind-Tunnel Drives," *Transactions AIEE,* March 1942.
10. R. F. Woll, "Unbalanced Voltage and Wound-Rotor Motors," *Westinghouse Engineer,* May 1944.
11. E. J. Posselt, "Speed Control and Braking for A.C. Cranes," *Iron and Steel Engineer,* May 1944.
12. P. B. Harwood, "Controllers for Wound-Rotor Motors," *Electrical Contracting,* October–November 1944.
13. N. L. Schmitz, "Control of Slip-Ring Motors by Means of Unbalanced Primary Voltages," *Proceedings AIEE* 1947, Section T7134.
14. Hal Gibson, "Wound Rotor Machine Operation," *General Electric Review,* April 1948.
15. B. W. Jones, "Unbalanced Resistance and Speed-Torque Curves," *General Electric Review,* September 1948.
16. E. L. Schwarz-Kast, "Synchronized Drives with Standard Electric Motors," *Machine Design,* April 1950.
17. H. L. Garbarino and E. T. B. Gross, "The Goerges Phenomenon—Induction Motors with Unbalanced Rotor Impedances," *Proceedings AIEE* 1950, Section T0255.
18. W. C. Carl, "Which Crane Hoist Control?" *Westinghouse Engineer,* March 1952.
19. W. R. Wickerham, "Load-O-Matic A-C Crane Hoist Control," *Westinghouse Engineer,* March 1952.
20. R. F. Woll, "Applying the Wound-Rotor Motor," *Westinghouse Engineer,* March 1953.
21. G. W. Heumann, Starters for Industrial Type A-C Motors," *Heating and Ventilating,* April 1953.

174 CONTROL OF WOUND-ROTOR MOTORS

22. O. T. Evans and L. Abram, "The Application of Dynamic Braking to A.C. Mine Winders," *The Metropolitan-Vickers Gazette*, September 1953.
23. S. J. Campbell, "Integral Horsepower Synchro Systems," *Electrical Manufacturing*, November 1953.
24. G. W. Heumann, "Types of A.C. Motor Controls, Part II," *Mill and Factory*, January 1955.
25. P. L. Alger and Y. H. Ku, "Speed Control of Induction Motors Using Saturable Reactors," *AIEE Paper* CP56-946, *Electrical Engineering*, February 1957.
26. H. A. Zollinger, "A.C. Crane Control with Load-O-Matic," *Westinghouse Engineer*, March 1957.
27. A. T. Bacheler, "A Comparison of Adjustable Frequency A.C. and Synchrotie Systems for Synchronized Drives," *AIEE Paper* CP58-808.
28. W. R. Harding, "Rectiflow Drives," *Westinghouse Engineer*, July 1958.
29. B. N. Garudacher and N. L. Schmitz, "Polyphase Induction Motors with Unbalanced Rotor Connections," *AIEE Paper* 58-1158.
30. W. Leonhard, "Elements of Reactor Controlled Reversible Induction Motor Drives," AIEE Paper 58-1176.
31. J. F. Szablya, "Torque and Speed Control of Induction Motors Using Saturable Reactors," *AIEE Paper* 58-1320.
32. L. R. Foote, "Adjustable Speed Control of A. C. Motors," *AIEE Paper* CP59-84.
33. W. Shepherd and G. R. Slemon, "Rotor Impedance Control of the Wound Rotor Induction Motor," *AIEE Paper* 59-138.
34. C. W. Chapman, "New Reactor Controlled A-C Crane Drives," *Proceedings of AIEE Material Handling Conference*, Milwaukee, Wisconsin, April 14–15, 1959.
35. H. A. Zollinger, "Application of Reactor Control to A-C Motors," *Westinghouse Engineer*, September 1959.
36. H. A. Zollinger, "Reactor Control of Induction Motors," *Electrical Manufacturing*, January 1960.
37. N. Onjanow, "A-C Drive Offers System Design Flexibility," *Electro-Technology*, December 1960.

PROBLEMS

1. A wound-rotor motor is rated 450 hp, 294 rpm, three-phase, 60 cycles, full-load secondary current 292 amp. Determine:

(a) Rheostat ohms and base secondary ohms.

(b) External secondary resistance per phase required to obtain a starting torque of 11,000 lb-ft.

(c) Speed at which the motor runs if an external resistance of 0.75 ohm is connected in each secondary phase and the motor is driving against a load torque of 6500 lb-ft.

(d) Secondary starting current and accelerating time when the motor is started with resistance as in (c), assuming that the load is a pure friction load and the Wk^2 of motor and load is 3900 lb-ft^2.

(e) External resistance needed to reduce the fan speed 10 per cent if the motor is driving a fan which, at full speed, requires a driving torque of 7500 lb-ft.

PROBLEMS

(*f*) External resistance needed, for plugging motor to a standstill, and limiting the initial plugging torque to approximately rated motor full-load torque.

2. A wound-rotor motor is rated 250 hp, with secondary full-load current of 185 amp; five secondary contactors are used for acceleration; the starting resistor is Class 115. Determine:

(*a*) Resistance of starting resistor sections, assuming equal current peaks and full load on the motor.

(*b*) Current-carrying capacity of resistor sections in terms of duty-cycle rating and equivalent continuous rating for grids.

(*c*) Size of secondary contactors.

3. The motor according to problem 2 is to be plugged.

(*a*) Determine the additional amount of resistance to be added as a plugging section, so that initial plugging torque does not exceed 1.25 per unit.

(*b*) Draw the necessary power and control circuits, using pneumatic timers to sequence secondary contactors.

4. Mark the following statements true or false, and give reasons:

(*a*) Current-limit control of wound-rotor motors limits the magnitude of accelerating current peaks.

(*b*) Current-limit control forces acceleration of motor regardless of load.

(*c*) Definite-time control of wound-rotor motors permits motor acceleration regardless of load.

5. A crane is equipped with motors having the following ratings:

Hoist: 100 hp, 1750 rpm, 30 min, 440 volts, three-phase, 60 cycles, 146 amp secondary, with mechanical load brake.

Trolley: 25 hp, 1730 rpm, 30 min, 440 volts, three-phase, 60 cycles, 68 amp secondary.

Bridge: 60 hp, 1740 rpm, 30 min, 440 volts, three-phase, 60 cycles, 116 amp secondary, with magnet holding brake, having torque equal to 75 per cent of rated motor torque.

(*a*) Select proper torque rating for bridge brake.

(*b*) State size of primary contactor, and number and size of secondary contactors, to be used on controllers.

6. The hoist controller according to problem 5 performs in accordance with Fig. 23-3. For a Class 162 resistor, determine:

(*a*) Resistance of each section.

(*b*) Current-carrying capacity of each section in terms of the Class 162 current rating of resistor units.

(*c*) Required continuous current rating of grid-type resistor sections.

7. An a-c crane bridge is equipped with two wound-rotor motors, each rated 125 hp, 220 volts, three-phase, 258 amp secondary. A duplex reversing plugging controller is used. Determine:

(*a*) Number of primary contactors used and their current rating.

(*b*) Number of secondary contactors used and their current rating.

(*c*) Number of overload relays used and how they should be set.

8. A crane hoist controller with single-phase lowering is equipped with a 75-hp motor, and 68 hp is required to hoist full load of 100 tons. Performance is according to Fig. 30-3. A load of 75 tons is being lowered. Mechanical efficiency of hoist is 85 per cent. What percentage of the kinetic energy of the load is absorbed electrically by stepping back from point 3-*L* to point 1-*L*?

4 CONTROL OF SYNCHRONOUS MOTORS

It has been stated that an induction motor can operate at synchronous speed only if a voltage is impressed on its rotor. As slip at synchronous speed is zero, frequency of the voltage impressed on the rotor is zero. In other words, direct current flows through the rotor winding. If direct current is maintained at sufficient strength, a flux is set up stationary with respect to the rotor. The poles of the rotor flux are attracted by opposite poles of the rotating flux set up by the stator winding. Thus the rotor rotates at the same speed with which the stator flux rotates in space, defined previously as the synchronous speed. It is determined by the number of poles (which must be equal for the stator and rotor) of the motor and the line frequency applied to the stator. In Table 7, synchronous speeds are listed for various numbers of poles and line frequencies.

SYNCHRONOUS MOTOR CHARACTERISTICS

For reasons of economy, the field structure of synchronous motors consists usually of a rotor with salient poles. As a matter of principle, either the stator or the rotor may be the d-c-excited field member. In practically all cases, however, the rotor is the field member and the stator is connected to the line. This arrangement has the advantage that the stator winding can easily be insulated for high voltages. Also, large currents do not have to be carried through slip rings. As compared with induction motors, synchronous motors have the following advantages:

1. Their efficiency is higher.

2. By proper control of the d-c field excitation, their power factor can be made unity or even leading, thereby improving the power factor of the industrial power system.

3. Mechanically, they have a simpler rotor construction, and their air gaps are larger.

SYNCHRONOUS MOTOR CHARACTERISTICS

Table 7. Synchronous Speeds for Various Numbers of Poles and Line Frequencies

Number of Poles	25 Cycles	Speed at 50 Cycles	60 Cycles
2	1500	3000	3600
4	750	1500	1800
6	500	1000	1200
8	375	750	900
10	300	600	720
12	250	500	600
14	214	428	514
16	188	375	450
18	166	333	400
20	150	300	360
22	136	273	327
24	125	250	300
26	115	231	277
28	107	214	257
30	100	200	240
32	94	188	225
36	83	167	200
40	75	150	180

Some source of direct current must be available. Either an external d-c shop system is used, or an exciter is directly coupled to or belt-driven from the synchronous motor shaft. Figure 1-4 shows cut-away views of high-speed and low-speed synchronous motors. The stators and their windings are constructed essentially like those of induction motors. The rotors contain laminated poles which carry the d-c-excited field coils which are terminated at slip rings. On the high-speed motor according to Fig. 1-4A, an exciter is mounted directly on the synchronous motor. The exciter rotor is assembled on the motor shaft, and the commutator brushes are connected to the brushes feeding the slip rings of the motor.

Direction of rotation of a synchronous motor depends upon the direction of rotation of the stator flux. Synchronous motors are reversed in the same way as induction motors, namely by reversing two of the stator leads.

Synchronous motors are constant-speed motors, and the only pos-

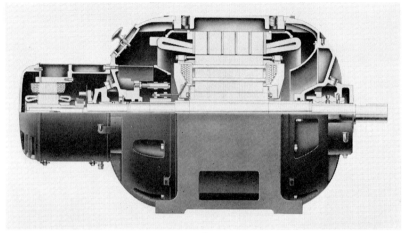

A

B

Fig. 1-4 Cut-away view of synchronous motors. *A*. High-speed motor with direct-connected exciter. *B*. Low-speed motor.

SYNCHRONOUS MOTOR CHARACTERISTICS

sible way to vary their speed is to vary the frequency of the voltage applied to the stator. Such a system of speed control would be impracticable except for very special cases. Because of the rigid speed characteristic that does not permit any speed change with load, special consideration must be given to synchronous motors operating in parallel on the same mechanical load, to insure proper division of load.

When the speed of a synchronous motor is called "constant," it must be understood that this expression pertains only to the average speed. The instantaneous speed does change with variations in load. Since the rotation of a synchronous motor depends upon the interaction of stator and rotor poles, it is possible that the angular relation between the axis of the stator poles and the axis of the rotor poles undergoes change with changes in torque, resulting in instantaneous speed changes.

There is a maximum torque that the motor can deliver, called the "pull-out torque." If it is exceeded, the motor pulls out of step and stalls. Since torque is proportional to flux and stator current, pull-out torque is a direct function of the product of field current and line voltage.

A synchronous motor containing only a stator and a d-c-excited rotor winding cannot develop any significant starting torque. When synchronous motors were introduced, they were started at no load by auxiliary motors, and then synchronized to the line. To obtain starting torque, enabling the motor to accelerate to a sufficiently high speed to permit synchronizing the motor to the line, a squirrel-cage winding is now provided on the rotor. This winding consists of bars embedded in the pole faces and short-circuited by end rings. This winding is also called "damper" or "amortisseur" winding, because it has the additional effect of damping instantaneous speed oscillations upon sudden variations in load.

Synchronous motor operation involves the following control features:

1. Control of acceleration and retardation, which concerns primarily the operation of the motor as an induction motor with d-c field not excited.

2. Application of d-c field excitation, after the motor has accelerated to a speed slightly below synchronous speed, to effect automatic synchronizing.

3. Control of the d-c field excitation, after synchronization has taken place, to obtain specific operating performance.

STARTING AS A SQUIRREL-CAGE MOTOR

The starting characteristics of synchronous motors with squirrel-cage winding are essentially the same as those of squirrel-cage induction motors. Starting torque and starting current depend largely upon the resistance of the squirrel-cage winding, which can be influenced by the selection of cross-section and material of the bars. Considerable latitude of choice is left to the motor designer. A low-resistance squirrel-cage winding produces a low starting torque and a high torque near synchronous speed, resulting in a low slip at which the motor pulls into synchronism. Also, a low-resistance squirrel-cage winding is more effective as a damper winding in suppressing torque oscillations. A high-resistance squirrel cage produces high starting torque which is necessary for many applications. On the other hand, the slip from which the motor is required to synchronize is increased, and the effectiveness of the squirrel cage as a damper winding is impaired. Generally speaking, high-speed motors of a given rating have higher starting torque than low-speed motors. Shown in Fig. 2-4 are typical

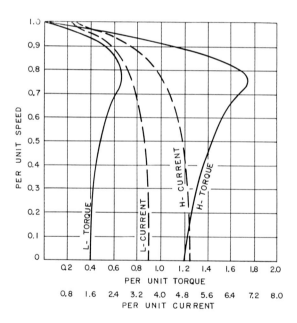

Fig. 2-4 Starting torque and current of typical high-speed and low-speed synchronous motors. Curves H = high-speed motor; curves L = low-speed motor.

STARTING AS A SQUIRREL-CAGE MOTOR 181

starting torque and current curves for high-speed and low-speed synchronous motors.

If the power system permits, synchronous motors can be started by applying full voltage to the stator terminals in the same manner as described for full-voltage starting of squirrel-cage motors. If system requirements demand a reduction in starting current, reduced-voltage starting and, in some special instances, part-winding starting may be used, as for squirrel-cage motors. When reduced-voltage starting is employed, motor torque varies as the square of the ratio of applied terminal voltage to normal voltage. Part-winding starting is generally limited to large slow-speed motors. Stator windings are arranged in two or three sections permitting part-winding starting in two or three steps. Since these motors are of special design, starting torque and starting current will have to be obtained from the motor designer.

After the motor has attained synchronous speed, the squirrel-cage winding does not carry current except when it acts as a damper winding, as when the axis of the rotor flux shifts with respect to the axis of the stator flux under the impact of sudden load changes. Squirrel-cage windings are not designed to carry starting current continuously. Precautions must be taken to prevent overheating of the squirrel-cage winding in the event the motor should fail to synchronize and attempt to run as an induction motor for a prolonged period of time. The permissible stall time of synchronous motors varies considerably, depending on the specific motor design. In most cases, the permissible stall time of synchronous motors is considerably shorter than the permissible stall time of squirrel-cage motors of equivalent rating.

While the motor is accelerating, the field winding of the rotor intersects the stator flux. As the field winding consists of a large number of turns, high voltages of the order of 10,000 volts or more might be induced in the field winding, if it were left open-circuited.

During the starting period, the field winding should be connected to a discharge resistor so that current may flow in it. The voltage across the field terminals is thereby limited to the product of induced field current and discharge resistance. The proper amount of discharge resistance to be used depends on the design of the field structure and should be obtained from the motor designer. Short-circuiting of the field winding results in a reduction of the available starting torque. When particularly heavy starting requirements must be met, it may be necessary to leave the field winding open-circuited. If this is the case,

182 CONTROL OF SYNCHRONOUS MOTORS

the field winding must be sectionalized, to keep the voltage induced in each section at a safe value.

For extra-heavy starting duty, synchronous motors have been built with a distributed starting winding, similar to the rotor winding of a wound-rotor induction motor. This winding is connected to slip rings, and the motor can be started like a wound-rotor motor, with external resistance connected to the starting winding slip rings.

PLUGGING AND DYNAMIC BRAKING

On some applications, such as rubber mills, it is necessary to bring the motor to a rapid stop within a few revolutions. One method is to remove field excitation and to reverse two stator phases. The motor is plugged, just like an induction motor, and its squirrel-cage winding produces plugging torque. This method of retardation is simple, but the amount of plugging torque available is limited and cannot be varied by external means. In the control, steps must be taken to prevent reversing the motor by disconnecting it from the line when it has come to a stop.

Another method to obtain rapid retardation is by dynamic braking. The motor is disconnected from the line and the stator winding is connected to an external resistor, field excitation remaining established. The motor then acts as a synchronous alternator, driven by the load inertia, and the kinetic energy of the motor rotor and the driven machinery is dissipated in the dynamic braking resistor. No reversing takes place when the motor comes to a standstill.

The dynamic braking torque varies with the amount of dynamic braking resistance. A maximum amount of average braking torque, that is a minimum number of revolutions between full speed and standstill, is obtained with a dynamic braking resistance of

$$R_{DB} = \frac{X_{DT}}{\sqrt{3}} \text{ ohms} \qquad (27) \bullet$$

X_{DT} is the transient reactance of the motor in the direct axis. This value is not apparent from the usually published motor data and must be obtained from the motor designer.

SYNCHRONIZING

After a synchronous motor has been started and accelerated to the maximum speed (or minimum slip) it is capable of attaining as an induction motor, field excitation is applied by connecting the field

SYNCHRONIZING

winding to a source of direct current. A synchronizing torque is then developed which must overcome the load torque, the torque corresponding to the losses of the motor out of synchronism, and accelerate the motor and the load from minimum slip to synchronous speed. Depending upon the condition of the load and the design of the motor, there is a certain amount of slip which must not be exceeded, or the motor will fail to synchronize.

If the load torque is small, as in a motor-generator set where the only load consists of friction and windage of the generator, it is not very important at which point in the starting cycle field excitation is applied. If only sufficient time is allowed for the motor to come up to full induction-motor speed, field may be applied without any particular precautions, and the motor will pull in step. However, if the motor has to synchronize against an appreciable amount of load torque, certain precautions have to be taken to apply field at the most advantageous point in the starting cycle.

For an explanation of the performance of the motor near synchronous speed, it is necessary to consider the effect of the saliency of the rotor. Since the rotor is not a smooth cylinder but carries salient poles, the reluctance of the rotor path to the stator flux varies. When the axis of the stator flux lines up with the rotor poles, the reluctance is low. When the axis of the stator flux falls in between the rotor poles, the reluctance is high. Variations in reluctance cause the reactance of the motor to vary, depending on the relative position of the rotor with respect to the stator flux. As the rotor slips with respect to the rotating stator flux, the current, power factor, and torque vary periodically, and instantaneous values deviate from the average value.

Figure 3-4 indicates the effect of rotor saliency. Upper sketch A is a vector diagram of line voltage and current during starting. The locked-rotor current, that is the starting current at the instant voltage is applied to the stator, is large and lags the line voltage by about 75 degrees. The circle at the end of the current vector is the locus of the current vector tip, representing the periodic current variations, both in magnitude and in power factor, due to the influence of varying reactance because of rotor saliency. As the motor accelerates, the magnitude of the current decreases and the power factor increases, as indicated by the dotted current vectors, until minimum slip is reached. The upper solid current vector represents this condition. For purposes of comparison, the full-load current corresponding to operation at synchronous speed is also indicated. It leads line voltage by an angle which depends on the magnitude of field excitation.

CONTROL OF SYNCHRONOUS MOTORS

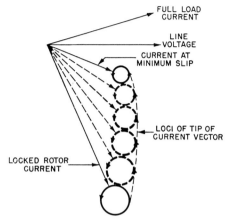

A — VECTOR DIAGRAM OF CURRENT DURING STARTING

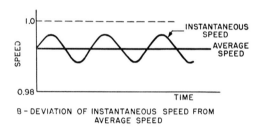

B — DEVIATION OF INSTANTANEOUS SPEED FROM AVERAGE SPEED

Fig. 3-4 Effect of saliency on starting performance of synchronous motor.

The actual speed which the motor attains at the end of its starting cycle as an induction motor is indicated in Fig. 3-4B. Because of variations in current and torque due to saliency, instantaneous speed varies periodically above and below average speed at twice slip frequency. If field is applied while the motor accelerates from a point of minimum speed to a point of maximum speed within a speed cycle, a lesser amount of synchronizing torque is required to accelerate the motor to synchronous speed. Conversely, while the motor is decelerating from a point of maximum speed to a point of minimum speed, a greater amount of synchronizing torque is required for acceleration. Therefore, the exact instant at which field is applied has an effect on whether or not a given motor will synchronize under load.

While the motor is operating out of synchronism, the rotor slips with respect to the stator flux. The relative position of the rotor poles with

SYNCHRONIZING 185

respect to the poles of the stator flux changes periodically. The angle, in electrical degrees, between the rotor axis and the stator-flux axis at which excitation is applied to the motor has an effect on the amount of load torque against which the motor is capable of pulling into synchronism.

The effect of the relative position of the rotor poles with respect to the stator poles is indicated in Fig. 4-4. Two poles of the stator and rotor are shown, developed in a straight line. The direction of stator flux rotation and mechanical rotor rotation is to the right. The slip of the rotor, that is the relative movement of the rotor with respect to the stator flux, is to the left. Figure 4-4A shows the instant at which the axis of the stator poles coincides with the axis of the rotor poles. Flux linkages between stator and rotor field are maximum.

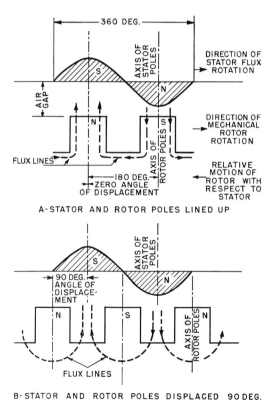

Fig. 4-4 Effect of displacement of synchronous motor rotor with respect to stator flux.

186 CONTROL OF SYNCHRONOUS MOTORS

Correspondingly, reluctance is low and reactance is high. Stator current is minimum, and voltage induced in the rotor field winding is zero. This position corresponds to the position the rotor assumes when the motor operates in synchronism with the field excited, except that the rotor lags the stator by a small angle determined by the load. Synchronous operation depends on high flux linkages between stator and rotor. In fact, a certain amount of synchronizing torque is developed in any salient pole machine which tends to hold the rotor in synchronism with no excitation applied to the rotor.

In conventional synchronous motors, this torque is small, and the motor will not stay in synchronism even when it is only lightly loaded. However, small motors, so-called "reluctance motors," are built which consist of a three-phase stator and a salient pole rotor with no field winding. Such motors have a low power factor and low efficiency. They are only practicable in sizes up to 1 horsepower and are generally used as auxiliary motors, for instance in regulating systems, when it is desired to obtain a mechanical speed which is synchronous with respect to an alternating-current frequency. Reluctance motors are controlled simply by full-voltage starters of the same type as are used for starting squirrel-cage motors. Their use in industry is rather limited.

Figure 4-4B shows the instant at which the rotor has slipped back 90 degrees. Flux linkages between stator and rotor field are minimum because the reluctance offered to the flux is maximum. Stator current is maximum, and current induced in the rotor field winding is maximum in the positive direction, namely in a direction to maintain the flux.

A criterion for the most favorable angle of rotor position at which field should be applied is based on the following considerations. If field is applied in a rotor position which obtains a condition of flux linkages between stator and rotor field similar to the one obtained during synchronous operation, available torque for synchronizing motor and load is highest. The most favorable angle is that at which the rotor lags the stator flux by a small amount. In that position, flux linkages are high, and voltage induced in the rotor field winding is just rising from zero in a positive direction. The most unfavorable angle is somewhat beyond the 90-degree displacement position. Flux linkages are low, and induced rotor current has just attained its maximum value in a positive direction.

This condition is indicated in Fig. 5-4. Load torque against which a given motor will synchronize, expressed in per unit of full-load torque, is plotted versus the angle of rotor displacement at which field excitation is applied. There is a distinctly favorable and an unfavorable

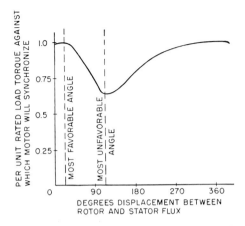

Fig. 5-4 Effect of angle at which field is applied on pull-in torque of synchronous motor.

range. The exact location of the most favorable and the most unfavorable angles changes as much as 60 degrees with different motors, and so does the amount of available pull-in torque. However, the general shape of the curve shown is typical.

OPERATION UNDER OVERLOAD

Each synchronous motor has a rated full-load torque, a pull-in torque at which it synchronizes, and a pull-out torque at which it loses synchronism or falls out of step. At rated full-load torque, the rotor poles lag the stator poles by an angle, called the "load angle." If the load torque increases, the load angle increases. At a load angle of approximately 90 degrees or somewhat less, the torque of the motor reaches a maximum value, called the "pull-out torque." If the load torque increases beyond that amount, the load angle is further increased, but the motor torque decreases. As a result, the rotor slips relative to the stator flux, the motor loses synchronism, it decelerates and stalls. If the field remains connected to the d-c source, torque pulsations occur, which at standstill may exceed three times rated torque. The stator current rises considerably, and the power factor becomes lagging. To prevent damage to the motor and driven machinery, field excitation should be removed quickly when a synchronous motor falls out of step.

Torque pulsations during operation out of synchronism with the field excited place motor windings under additional stress, which may

188 CONTROL OF SYNCHRONOUS MOTORS

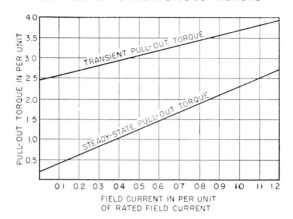

Fig. 6-4 Pull-out torque of a typical synchronous motor.

eventually shorten the life of the motor. The greatest degree of protection of the motor against the effects of these pulsating torques is obtained by removing the field immediately as the motor falls out of step, before it has slipped a pole. If pull-outs do not occur frequently, slipping one or two poles is tolerable before field excitation is removed.

Whether or not the motor should also be disconnected from the line depends on the character of the load. If the motor is pulled out of step because of a mechanical overload, and if the overload is likely to be of short duration, the motor may remain connected to the line with field excitation removed. It will continue to operate as an induction motor. If the load torque falls below the induction-motor torque developed by the synchronous motor at the maximum slip from which it will synchronize, it may go through a resynchronizing cycle. While the motor operates as an induction motor, care must be taken to protect the squirrel-cage winding on the rotor against overheating, and the motor must be disconnected from the line if the squirrel-cage winding reaches its permissible temperature rise.

The pull-out torque of a given motor varies with its field excitation. It increases as the field current is increased. The lower curve of Fig. 6-4, represents the steady-state pull-out characteristic of a typical synchronous motor, in relation to field current. Pull-out torque, as given by that curve, is obtained when the load torque is raised gradually up to the point at which the motor loses synchronism.

If an overload is applied suddenly, the motor develops a substantially higher pull-out torque than indicated by the steady-state pull-out

POWER-FACTOR CORRECTION

characteristic, but only for a very short time. The so-called "transient pull-out torque," or maximum instantaneous torque peak the motor can tolerate without pulling out of step, is given by the upper curve in Fig. 6-4. Transient pull-out torque is caused by the inherent sluggishness of the magnetic circuit of the motor. If the load is increased, the demagnetizing action of the stator load current reduces the flux linkages between the stator and the rotor, causing the angle of displacement between rotor and stator poles to increase. This action cannot take place instantly. Thus the motor develops a higher pull-out torque than the steady-state value, until the demagnetizing action of the stator load current has taken effect. The length of time during which the motor is capable of producing transient pull-out torque is about a second or a substantial fraction thereof.

POWER-FACTOR CORRECTION

Stator current and power factor of a synchronous motor vary with field excitation. In Fig. 7-4, the stator current is plotted versus field current for various amounts of load. Because they are shaped like the letter V, these curves are called "V-curves." For a given load, the stator current is minimum and power factor is unity at a certain field current. If the field current is increased, the power factor is leading;

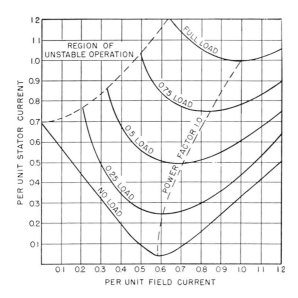

Fig. 7-4 V-curves of a typical synchronous motor rated at 1.0 power factor.

CONTROL OF SYNCHRONOUS MOTORS

if the field current is decreased, the power factor is lagging. The broken line in Fig. 7-4 connects points of unity power factor. In the region to the right, the power factor is leading; in the region to the left, the power factor is lagging.

Standard synchronous motors are usually rated to carry full load at unity power factor or at 0.8 leading power factor. Synchronous motors can be used not only for driving a mechanical load, but also to generate reactive kva which improve the power factor of the supply system. If the mechanical load on the motor is not constant, the excitation system can be arranged so that field current varies as a function of load and that the motor either operates over its load range at a constant power factor or supplies constant reactive power to the supply system.

CONTROLLERS FOR SYNCHRONOUS MOTORS

Controllers for synchronous motors perform two functions. One is to start the motor from the a-c line as a squirrel-cage motor. Depending on the load or on power-system limitations, the a-c starting portion of the controller may contain the equivalent of a full-voltage starter, a reduced-voltage starter, a part-winding starter, or a wye-delta starter.

The second function of synchronous motor controllers is the application of d-c excitation to the motor field. Provision also has to be made for removing field excitation when the motor pulls out of synchronism, and the squirrel-cage winding must be protected against overheating while the motor is operating out of synchronism. In addition to a-c starting equipment, a conventional general-purpose synchronous-motor controller includes:

1. A field contactor which closes to connect the field winding to a d-c source and, when dropped out, connects the field winding to a discharge resistor.

2. Means for energizing the field contactor at the correct point of the starting cycle, and for removing excitation when the motor pulls out of step.

3. A field-discharge resistor which prevents excessive voltage from appearing across the field winding when d-c excitation is removed.

Optional items included in many synchronous motor controllers are:

Field rheostat or tapped field resistor for adjustment of the field current.

FIELD APPLICATION BY TIMING RELAY

Line ammeter and field ammeter.

Protective relay to prevent overheating of the squirrel-cage winding when operating out of synchronism.

Various systems have been developed to control the closing of the field contactor. In the following sections, several methods of field application are described which are in practical use. For the sake of simplicity, all connection diagrams are drawn for full-voltage starting of a low-voltage motor. Any other a-c starting method may be substituted without affecting the field control circuits.

When field excitation is obtained from a constant-voltage d-c source, independent of the motor, the field resistor or rheostat is connected in series with the motor field winding. When an individual exciter is used which is driven by the motor, the rheostat is connected in the exciter field circuit, and the motor field current is adjusted by varying the exciter voltage. During starting, the exciter field rheostat is short-circuited in most cases to obtain a faster buildup of the exciter voltage.

FIELD APPLICATION BY TIMING RELAY

Starting a synchronous motor with no load or with a light load, which occurs for instance on motor-generator sets, does not require any special precautions in regard to the rotor position, relative to the stator, at which field excitation is applied. It suffices to let the motor accelerate as an induction motor to minimum slip, since sufficient synchronizing torque is developed at any rotor position to pull the motor into synchronism. A very simple control system is obtained if the closing of the field contactor is made a function of the time elapsed after the closing of the line contactor, as long as sufficient time is allowed to accelerate the motor to minimum slip.

The principal circuits of a full-voltage starter with definite-time field application are shown in Fig. 8-4. After the motor has been connected to the line, an interlock on line contactor M energizes a motor-driven timing relay FR, the time-closing contact of which energizes field contactor FC, which connects the motor field to the d-c source and disconnects the field-discharge resistor. The circuits shown here do not include field removal after the motor pulls out of step, nor protection of the squirrel-cage winding against overheating due to operation out of synchronism. Several methods for accomplishing these functions are described in subsequent paragraphs, and they may be added to the field-application circuits.

CONTROL OF SYNCHRONOUS MOTORS

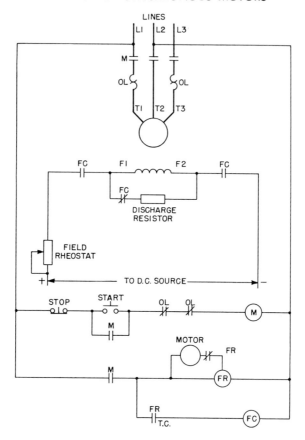

Fig. 8-4 Connections of synchronous motor starter with field application by definite-time relay. M = line contactor; FC = field contactor; FR = definite-time field relay; OL = overload relays.

FIELD CONTROL BY POLARIZED FIELD FREQUENCY RELAY

When the stator of a synchronous motor is connected to the line, the voltage induced in the field winding can be used to determine the proper instant at which to apply d-c field excitation. The frequency of the induced voltage is equal to the slip frequency. At standstill, it equals line frequency. As the motor accelerates, the frequency drops, becoming zero at synchronism. The change in slip frequency can be utilized as a signal to apply field excitation.

Figure 9-4 shows the principal circuits of a synchronous motor starter utilizing a polarized field frequency relay for the control of applica-

FIELD CONTROL BY POLARIZED FIELD FREQUENCY RELAY   193

tion and removal of the field excitation. This system, originally developed by the Electric Machinery Manufacturing Company, is also used by such control manufacturers as Allen-Bradley Company, Allis-Chalmers Manufacturing Company, Clark Controller Company, Cutler-Hammer, Incorporated, Ideal Electric Company, and Square D Company.

Field relay FR performs a dual function: it controls the application of field excitation during starting, and it also removes field in case the motor should pull out of step. Figure 10-4A is a view of this relay,

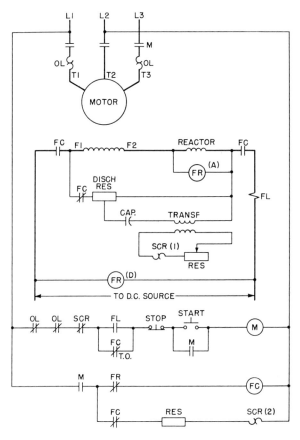

Fig. 9-4 Connections of synchronous motor starter with field control by polarized field frequency relay. M = line contactor; FC = field contactor; FR = polarized field frequency relay; FL = field loss relay; OL = overload relays; SCR = squirrel-cage protective relay.

194 CONTROL OF SYNCHRONOUS MOTORS

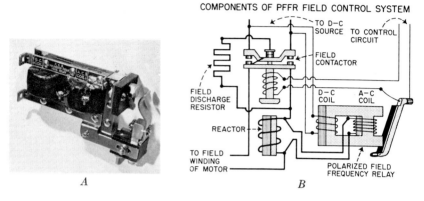

Fig. 10-4 Polarized field frequency relay. *A*. View of relay. *B*. Schematic diagram of components of field-control system. (Courtesy Electric Machinery Manufacturing Company.)

whereas Fig. 10-4*B* illustrates the function of the relay in a schematic manner. The circuits are the same as those shown in Fig. 9-4, but the physical relationship of the components comprising the field-control system is also illustrated. Relay *FR* has two coils on a common magnetic structure which actuates a single armature. Coil *A* is an a-c coil which is connected in parallel with a reactor and in series with the motor field winding. The current flowing through this relay coil is alternating current of a frequency equal to the slip frequency of the voltage induced in the field winding. Coil *D* is a d-c coil which is connected across the d-c source supplying field excitation. The direct current flowing through this coil polarizes relay *FR*. The total flux in the magnet structure of the relay, which holds its armature closed, is the sum of the polarizing d-c flux and the a-c flux due to the voltage induced in the field winding. Because of the unidirectional flux supplied by the d-c coil, the total flux during positive half cycles of the a-c current wave is the difference between the a-c and d-c flux, whereas during negative half cycles the total flux is the sum of the a-c and d-c flux. In Fig. 11-4, the solid line represents the flux due to the induced a-c current. The broken line represents the resultant flux, that is the unidirectional flux supplied by coil *D* on which the a-c flux supplied by coil *A* is superimposed.

When line contactor *M* closes, the voltage induced in the motor field winding is at line frequency, and the reactance of the reactor is high. Current flowing through coil *A* of *FR* is high. The high resultant

FIELD CONTROL BY POLARIZED FIELD FREQUENCY RELAY 195

flux in the relay snaps the armature closed and opens the relay contact before field contactor FC has a chance to close. As the motor accelerates, slip frequency decreases and correspondingly the reactance of the reactor decreases, causing a reduction in the a-c current flowing through coil A of relay FR. At approximately 95 to 96 per cent of synchronous speed, the total flux during the positive half slip cycle becomes so small that relay FR drops out and closes its contact. Field contactor FC is energized and closes its contacts, applying excitation to the motor field. Referring to Fig. 11-4, it will be noted that relay FR has dropped out while the current induced in coil A was passing from a positive to a negative value and rising in a negative direction when the relay armature dropped out. At that time, the current induced in the motor field was changing in the same direction. Because of a slight time delay in the closing of the relay contact, due to movement of the armature and the additional time required for field contactor FC to close, excitation is actually applied to the motor field winding while the current induced in the field winding is changing in a positive direction. This means that field application occurs in a range of favorable angle of displacement between stator flux and rotor poles.

While the motor is operating in synchronism, the polarizing excitation on coil D of relay FR remains established, but it is of insufficient magnitude to cause the relay armature to pick up. However, when the motor falls out of synchronism, an a-c voltage is induced in the motor field winding, and alternating current will flow through coil A of FR. When the a-c flux adds to the d-c flux during a nega-

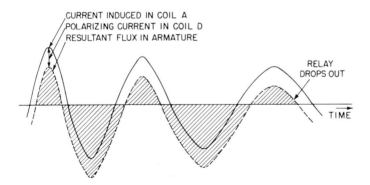

Fig. 11-4 Coil current and armature flux in polarized field frequency relay. (Courtesy Cutler-Hammer, Inc.)

196 CONTROL OF SYNCHRONOUS MOTORS

tive slip cycle, the total flux causes the armature of FR to pick up. The contact of FR opens and drops out contactor FC, thereby removing d-c field excitation. The motor remains connected to the line, operating as an induction motor. If the load drops and permits the motor to accelerate to minimum slip again, the motor may resynchronize. If the load remains too high and prevents reacceleration of the motor, squirrel-cage protective relay SCR will drop out line contactor M and shut down the motor before the squirrel cage is overheated. The function of the SCR relay is described later on under the heading, "Protection of Squirrel-Cage Winding."

In the circuit according to Fig. 9-4, a current relay FL is indicated. The coil of this relay is connected in the field excitation circuit. The relay drops out the line contactor and shuts down the motor if the field excitation circuit is open-circuited, resulting in loss of field. This relay is an optional feature which may be provided as additional protection of the motor against operation out of synchronism because of loss of field.

FIELD APPLICATION BY SLIP-FREQUENCY RELAY

Another method of utilizing the voltage induced in the motor field winding during starting as a criterion for field application is the use of a slip-frequency relay. Connections of a synchronous motor starter utilizing the slip-frequency relay are indicated in Fig. 12-4. Slip-frequency field relay FR is a magnetic flux-decay relay. It has two coils which are copper-jacketed. Closing coil CC is connected across the d-c field excitation source and closes the relay armature initially; synchronizing coil SYN is connected across a portion of the field-discharge resistor. Hence a-c current flows through the SYN coil of relay FR at slip frequency. A rectifier in series with the relay coil suppresses negative half cycles so that only positive half-wave current pulses flow through the relay coil, as indicated in Fig. 13-4. As long as the slip frequency is high, the current pulses are close together. The relay stays picked up between half cycles. As the time interval between half cycles increases progressively, a point is reached at which the period between half cycles becomes longer than the adjustable drop-out time of the relay. Thus the relay recognizes that, when it drops out, the slip frequency has dropped to a sufficiently low value to permit closing of the field contactor. The relay can be adjusted to drop out when the motor has attained a speed between 95 and 98 per cent of synchronous speed.

When the rectifier is connected as shown, positive half cycles of

FIELD APPLICATION BY SLIP-FREQUENCY RELAY 197

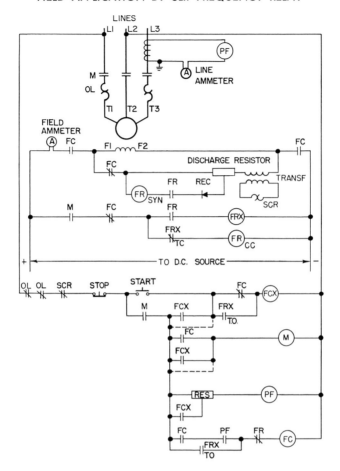

Fig. 12-4 Connection of synchronous motor starter with field application by slip-frequency relay. M = line contactor; FC = field contactor; FCX = auxiliary relay for FC; FR = slip-cycle field relay; FRX = auxiliary relay for FR; OL = overload relays; SCR = squirrel-cage protective relay; PF = power-factor relay.

current pass through the relay coil. The negative half cycles of induced a-c voltage across the relay coil are suppressed, and no current flows through the relay coil during negative half cycles. Therefore relay FR drops out close to the end of a negative half cycle and energizes field contactor FC.

Field contactor FC has an a-c magnet, and its closing time is short. Therefore field application occurs as the induced current in the motor

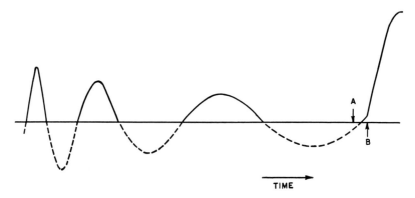

Fig. 13-4 Current through SYN coil of field relay FR prior to synchronizing.

field winding is just rising in a positive direction. This condition corresponds to a displacement between stator flux and rotor somewhat beyond 180 degrees. Hence field application takes place within the most favorable region of angular displacement.

Figure 14-4 shows a field control panel which is a subpanel containing the components required for application and removal of field excitation. This subpanel, together with the field contactor, are assembled on the synchronous motor controller in addition to the conventional a-c power-circuit devices required for starting the motor. In addition to field relay FR, the field panel contains these devices: auxiliary a-c relay FCX, which performs various auxiliary functions during the starting period of the motor; auxiliary d-c relay FRX, which acts as a voltage-check relay to permit field application and synchronization only if the voltage of the d-c source is at its proper level; and power-factor relay PF, which removes field and may disconnect the motor from the line during the first slip cycle if the motor pulls out of step. The function of the power-factor relay is described later in detail under the heading, "Field Removal under Overload."

Squirrel-cage protective relay SCR is a thermal overload relay which protects the squirrel-cage winding against overheating due to prolonged operation out of synchronism. Its function will be described in detail later on under the heading "Protection of Squirrel-Cage Windings."

The starting control sequence is as follows: pressing the start pushbutton energizes relay FCX, which in turn energizes line contactor M and recalibrates the voltage coil of relay PF. If the motor has a di-

FIELD APPLICATION BY SLIP-FREQUENCY RELAY

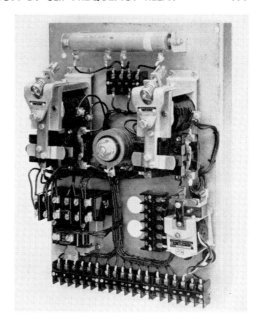

Fig. 14-4 Field control subpanel containing field control components for synchronous motor controller with field application by slip-frequency relay.

rect-connected exciter, FCX short-circuits the exciter field rheostat. If the motor drives a compressor, FCX may be utilized to actuate an automatic unloader. If the motor is equipped with a forced-oil lubrication system, FCX may bypass the oil-pressure interlocking devices until after the motor and the oil pump have come up to speed. The motor is now connected to the line, and the field winding is connected to the discharge resistor. The motor is accelerating as an induction motor, and an induced a-c current is flowing through the field and the discharge resistor. A portion of this current flows through the rectifier and the SYN coil of relay FR. This coil, however, does not develop sufficient pull to close the relay, although it is strong enough to hold that relay closed once it has been picked up.

If the d-c control circuit is connected to an independent d-c power source, the subsequent relay sequence will take place immediately. If the motor depends on the directly connected exciter for its source of d-c power, a time interval will elapse to permit buildup of exciter voltage. An interlock on M energizes closing coil CC of relay FR, which closes its armature. When the voltage of the d-c circuit is at its prescribed value, relay FRX picks up through an interlock on FR. One contact on FRX disconnects the CC coil of FR, so that this relay

200   CONTROL OF SYNCHRONOUS MOTORS

is now entirely under the control of its SYN coil. Other contacts on FRX prepare circuits for energizing the field contactor.

When the motor has accelerated to minimum slip, and the duration of one half slip cycle has become larger than the time adjustment of FR, that relay drops out at a point of the slip cycle indicated as A in Fig. 13-4. A normally closed contact on FR energizes contactor FC, which closes at point B in Fig. 13-4. The field winding is connected to the d-c power source and disconnected from the discharge resistor. The motor now pulls in step and operates in synchronism. At the same time, relays FR and FRX are disconnected from the d-c control circuit.

FRX is also of the magnetic flux-decay type, and it drops out after a time delay to allow any current surges due to synchronizing to subside. The opening of FRX contacts places FC directly under the control of relay PF, whose function is described in the next section of this chapter. It also causes relay FCX to drop out, thereby placing PF on its regular operating characteristic. M is now placed under the control of an interlock on FC. If FCX has been used to short-circuit the exciter field rheostat, to actuate a compressor bypass, or to bypass an oil-pressure interlock, these circuits are restored to their normal condition.

Now that the motor is operating in synchronism, the operator may adjust field excitation to a value corresponding to the desired power factor by observing the field and line ammeters. For this adjustment, he uses either the exciter field rheostat or a rheostat in series with the motor field.

Pressing of the stop pushbutton or tripping of the overload or the squirrel-cage protective relays deenergizes the line and field contactors, thereby shutting down the motor. Circuit conditions are restored so that the control may go through another starting sequence.

Another application of the slip-frequency relay is the Slipsyn * synchronous motor control of the Westinghouse Electric Corporation. Principal circuits of a full-voltage starter are indicated in Fig. 15-4. Field relay FR is a magnetic flux-decay relay having a slight time delay. It is equipped with two coils: a closing coil CC, and a synchronizing coil SYN. When the start pushbutton is pressed, line contactor M closes and energizes the motor, which accelerates as an induction motor. The motor field is connected across the discharge resistor, through the squirrel-cage protective relay SCR. At the same time field relay FR is picked up through its closing coil, and

---

* Registered trade mark of the Westinghouse Electric Corporation.

FIELD APPLICATION BY SLIP-FREQUENCY RELAY 201

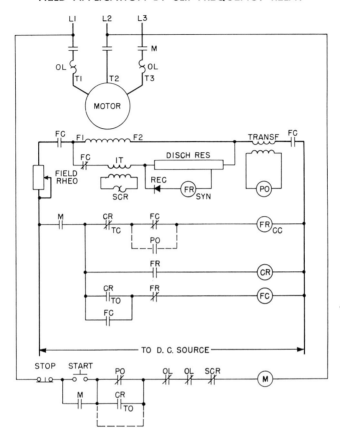

Fig. 15-4 Connections of Slipsyn synchronous motor controller with field application by slip-frequency relay. M = line contactor; FC = field contactor; FR = field-application relay; CR = auxiliary control relay; PO = pull-out relay; OL = overload relays; SCR = squirrel-cage protective relay; $1T$, $2T$ = transformers. (Courtesy Westinghouse Electric Corporation.)

auxiliary relay CR is energized. The purpose of relay CR is to set up the necessary circuits for the control sequence. Since it is a flux-decay drop-out relay, it provides a slight time delay when it drops out to prepare circuits for restarting, so that transients occurring in the field circuit during synchronizing do not disturb the control sequence.

When CR picks up, it disconnects the closing coil of FR. The field relay is now under the control of its synchronizing coil, which is connected across the discharge resistor in series with a half-wave rectifier, so that its coil current consists of unidirectional pulses at slip frequency. When the distance between current pulses becomes larger than the relay setting, indicating that the motor has accelerated to between 94 and 98 per cent of synchronous speed, FR drops out and energizes field contactor FC, which is equipped with a d-c magnet. At the same time, CR is dropped out. A pull-out relay PO is connected in the motor field circuit. The function of this relay is described in the following section on "Field Removal under Overload." Relay PO may respond to field-current transients while the motor pulls into step. CR nullifies this relay during synchronizing. Pressing of the stop pushbutton or opening of one of the protective relays drops out the line and field contactors. The equipment is then ready for another start.

FIELD REMOVAL UNDER OVERLOAD

When the motor falls out of step because of overload, it is desirable to remove field excitation as quickly as possible to avoid the mechanical stresses due to torque pulsations. When the motor loses synchronism, a-c current is induced in the field winding because of the slip of the rotor with respect to the stator flux. In the stator, the magnitude and power factor of the a-c current undergo significant changes. Either of these phenomena can be utilized for actuating a relay which causes field excitation to be removed.

An example of pull-out protection responding to a-c current induced in the field winding is the polarized field-frequency relay of Fig. 9-4. While the motor operates in synchronism, very little d-c current flows through coil A of relay FR. However, when a-c current flows in the field winding as the rotor pulls out of step, the reactance of the reactor increases, and a proportionately higher amount of a-c current is shunted through the relay coil. When the a-c current is in such a direction as to add to the polarizing flux of coil D, FR picks up and opens its contact, dropping out field contactor FC. When a field loss relay FL is included, it will drop out quickly and cause the motor to be disconnected from the line. When relay FL is not included, line contactor M remains closed. If the overload is removed, permitting the motor to accelerate again to minimum slip, the motor will resynchronize in the same manner as during a start from rest. If the overload persists and the motor does not accelerate again, the squirrel-cage protective relay SCR will shut down the motor before the squirrel cage is damaged.

In a similar manner, pull-out relay PO of Fig. 15-4 functions in re-

FIELD REMOVAL UNDER OVERLOAD

sponse to current induced in the field winding during a pull-out. A current transformer is connected in series with the motor field winding. The coil of relay PO is connected to the secondary of that current transformer. While the motor is in synchronism, direct current flows through the primary of the current transformer, and no voltage is induced in its secondary winding. Therefore no current flows in the relay coil, and its armature is dropped out. This relay has a very light armature, and it closes rapidly when a-c current at slip frequency flows in the relay coil. When the motor falls out of step, a voltage is induced in the secondary of the current transformer. Current now flows in the coil of relay PO, and its armature picks up. With connections as shown in Fig. 15-4, the opening of the PO contact causes line contactor M to open, thereby shutting down the motor and removing field excitation. If it is desired to allow the motor to resynchronize, the connections indicated by broken lines in Fig. 15-4 are added. Line contactor M is permitted to remain closed, whereas another contact on relay PO energizes the closing coil of field relay FR, thereby dropping out field contactor FC. Since the field circuit is now open-circuited, no current flows through the transformer, and relay PO drops out. Circuits are now set up for field relay FR to resynchronize the motor if the load conditions permit reacceleration to minimum slip. If the motor fails to resynchronize, relay SCR is relied upon to disconnect it from the line before the squirrel-cage winding overheats.

Response of pull-out protection to induced current in the motor field circuit permits the use of simple relays and simple circuitry. However, the motor must actually slip before the protective relay functions, thereby subjecting the motor to some extra stresses.

Another method of providing pull-out protection is to have the pull-out protective relay respond to the change in power factor of the motor stator current. What happens to the power factor when an overload is applied to the motor is illustrated in Fig. 16-4. As long as the motor develops its normal torque, the power factor is high. When overload is applied and the motor torque increases, the power factor decreases—slowly at first, then rapidly as pull-out occurs.

Power-factor relay PF of the controller shown in Fig. 12-4 responds to the phase angle of the motor armature current. This relay is a simple clapper-type relay, as illustrated in the lower right-hand corner of the field-control subpanel of Fig. 14-4. It has two coils. A voltage coil is connected across lines $L1$ and $L2$, and a current coil is connected to a current transformer in line $L3$.

204 CONTROL OF SYNCHRONOUS MOTORS

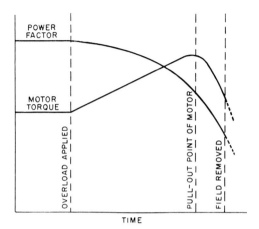

Fig. 16-4 Effect of overload on torque and power factor of synchronous motor.

Figure 17-4 illustrates the response of relay PF to power-factor shift during a pull-out. $E1$, $E2$, $E3$ represent the voltage vectors of the three-phase a-c line. V represents the voltage applied to the relay potential coil. I represents the motor current vector flowing in $L3$, which, at unity power factor, is in phase with voltage vector $E3$. The mmf acting on the relay armature is proportional to the vectorial sum of V and I.

The two circles, the center of which is in line with vector V, determine the conditions under which the relay is picked up or dropped out. Two circles are shown, since the mmf required to pick up the relay

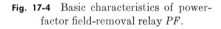

Fig. 17-4 Basic characteristics of power-factor field-removal relay PF.

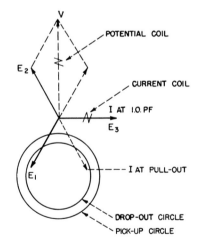

FIELD REMOVAL UNDER OVERLOAD

armature is larger than the mmf required to hold the armature closed. The relay is picked up when the end point of vector I stays outside the pick-up circle. The relay drops out when the end point of vector I falls inside the drop-out circle.

Normally, the motor operates at unity or leading power factor. In Fig. 17-4, the current vector would either coincide with voltage vector $E3$ or shift in a counterclockwise direction. In this region, the mmf of the power factor relay is sufficiently large so that the relay will be picked up and its contact closed.

When the motor is overloaded, its power factor becomes lagging, and the current vector shifts in a clockwise direction with respect to voltage vector $E3$. At the instant of pull-out, the current vector starts swinging in a clockwise direction; and during the first slip cycle out of synchronism, it rotates through 360 degrees, as the rotor slips one whole cycle with respect to the stator flux. When the end point of the current vector swings inside the drop-out cycle, relay PR drops out and opens its contact. Since this relay has a very light armature, relay drop-out always occurs during the early part of the first slip cycle. This means that pull-out protection functions before the first torque reversal has taken place, thus keeping extra stresses due to out-of-step operation of the motor, with the field energized, to a bare minimum.

Opening of the PF relay contact drops out field contactor FC, thereby removing field excitation. With connections as indicated by solid lines in Fig. 12-4, the opening of FC causes line contactor M to drop out, thereby shutting down the motor. Using the jumpers around contacts of relay FCX, as indicated by the broken lines, field only is removed, and the motor remains connected to the line. Field and control connections are reestablished as they were during the starting cycle. After the motor has reaccelerated to minimum slip, application of motor field and synchronization take place in the same manner as during the starting sequence.

Location and size of the pick-up and drop-out circles can be adjusted by adjusting the resistance in series with the voltage coil of relay PR, which determines the magnitude of coil voltage V, and hence the mmf due to the voltage-coil current. Increasing V increases the diameter of the circles and moves them downward. Consequently a larger current at lagging power factor is needed to drop out PF. Adjustment of the voltage-coil excitation should be such as to permit the motor to carry its maximum permissible overload without pulling out of step, and to assure that drop-out of the relay occurs during the first slip cycle. When an overload which barely misses causing a pull-

206 CONTROL OF SYNCHRONOUS MOTORS

out is suddenly removed, the rotor snaps back from a lagging to a leading load angle, and for an instant the motor acts like a generator. This shift in rotor position is accompanied by a wide swing of the current vector in a counterclockwise direction. Examination of Fig. 17-4 indicates that the current vector may swing through an angle of approximately 270 degrees, with no chance of its tripping the power-factor relay. This feature avoids inadvertent tripping due to wide current swings as a result of the sudden application and subsequent removal of impact loads.

At pull-out, the magnitude and phase angle of the current vector are similar to the condition existing when the motor operates at minimum slip during the starting cycle. In Fig. 18-4, the solid circles represent pick-up and drop-out performance of the power-factor relay when adjusted to obtain proper pull-out protection during operation in synchronism. The dashed current vectors represent motor current at standstill and at minimum slip, the dashed curve being the locus of the

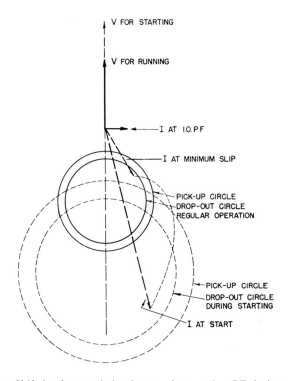

Fig. 18-4 Shift in characteristic of power-factor relay PF during starting.

PROTECTION OF SQUIRREL-CAGE WINDING 207

end point of the current vector during the starting cycle before synchronization takes place. The current vector at minimum slip may fall inside the relay drop-out circle, which means that it would be impossible to apply field because the PF relay contact would be open. To overcome this condition, relay PF is recalibrated during starting. A contact on FCX short-circuits a portion of the resistance in series with the PF voltage coil. As a result, voltage vector V (see Fig. 18-4) is increased to the dashed vector. Pick-up and drop-out circles are shifted. They are represented by the dashed circles. Relay pick-up at minimum slip is assured. After the motor has been synchronized, FCX drops out with a time delay (due to time-delay drop-out of FRX) to allow for any current surges to subside. The relay functions again according to the solid circles, and normal operation in synchronism ensues.

PROTECTION OF SQUIRREL-CAGE WINDING

While the motor operates in synchronism and at a constant load angle, no voltage is induced in the squirrel-cage winding, no current flows in it, and no heating takes place. However, during operation out of synchronism, an a-c current of slip frequency flows in the squirrel-cage winding, which causes heating. Since the squirrel-cage winding is designed for starting duty only, it has limited thermal capacity, which bears no relation to the thermal capacity of the stator winding. If prolonged operation out of synchronism occurs, either during a start against an overload or after a pull-out, the squirrel-cage winding is likely to heat much faster than the stator winding. In general, the thermal overload relay, which is selected to protect the stator winding, does not afford protection to the squirrel-cage winding.

One way of protecting the squirrel-cage winding is to use an incomplete-sequence relay, which causes the motor to be disconnected from the line if it operates out of synchronism for a certain length of

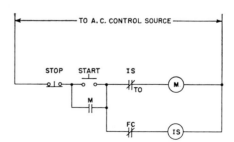

Fig. 19-4 Connections of incomplete-sequence relay. M = line contactor; FC = field contactor; IS = incomplete-sequence relay.

time. How an incomplete-sequence relay functions is illustrated in Fig. 19-4. When line contactor M is closed, an incomplete-sequence relay IS is energized through a normally closed interlock on field contactor FC. IS is a timing relay and it opens its contact after an adjustable time delay. Line contactor M is then dropped out. However, if field is applied by closing of FC before IS has timed out, the coil circuit of IS is opened and IS resets. The time which the motor may operate out of synchronism, either during starting or after a pull-out, is fixed irrespective of motor load condition, which has a bearing on the amount of current flowing in the squirrel-cage winding and which hence determines the heating of that winding. Therefore the incomplete-sequence relay time setting must be such as to prevent overheating of the squirrel-cage winding under the worst load condition.

The squirrel-cage protective relay SCR of Fig. 9-4 is a thermal relay which responds to the deflection of a bimetal strip due to heating caused by current. This relay has two elements. Element (1) is connected in the secondary of a transformer, which, in series with a capacitor, is connected in the circuit of the motor field-discharge resistor, thereby forming a frequency-sensitive network. When the motor stalls and fails to accelerate, a heavy current flows through element (1), causing SCR to heat up quickly and to trip in a short time. As the motor accelerates and slip frequency decreases, the current in (1) tapers off, thereby reducing the heating of SCR. Thus the heating of SCR reflects the effect of slip frequency, or motor speed, on the heating of the squirrel-cage winding. The thermal relay has a second element (2) which is energized from the control source as long as the field contactor FC is open. This element acts like a straight thermal timing relay, functioning like an incomplete-sequence relay. It shuts down the equipment if the motor fails to synchronize within a set period of time.

The SCR relays in Figs. 12-4 and 15-4 are thermal relays connected to the secondary of transformers in the motor field-discharge circuit. These relays respond to current and frequency flowing in the field-discharge circuit during motor operation out of synchronism. They reflect the actual heating of the squirrel-cage winding, since the amount of current which the SCR relay sees depends on slip frequency. Therefore a motor running out of synchronism is permitted to operate for a longer time than one which is stalled, that is to say, in which slip frequency equals line frequency. This reflects the fact that the squirrel-cage winding of a stalled motor heats much faster than that of a motor operating out of synchronism.

PROTECTION OF SQUIRREL-CAGE WINDING

A

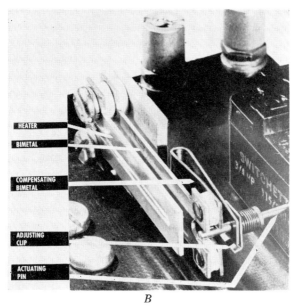

B

Fig. 20-4 Thermal-type squirrel-cage protective relay. *A*. Relay assembled on transformer. *B*. Close-up of thermal element.

210 CONTROL OF SYNCHRONOUS MOTORS

One form of SCR relay is illustrated in Fig. 20-4. A is a general view of the relay. The bimetal element and the contact unit are mounted on top of the transformer. B is a close-up of the bimetal element. A heater carries the current induced in the transformer secondary winding. This heater is in close proximity to a bimetal fork which, under the effect of the heat generated by the current in the heater, deflects and actuates a pin, which in turn operates the contact unit. Since bimetal elements respond to total temperature, the trip characteristic is affected by ambient temperature. To compensate for variations in ambient temperature, a compensating bimetal element is provided which shifts the point of contact between the bimetal and the actuating pin, thereby varying the distance through which the main bimetal element has to travel in order to actuate the contact unit.

SYNDUCTION MOTORS

Small synchronous motors of conventional design, say below 50 horsepower, are comparatively expensive. Reluctance motors are not very practicable in the integral horsepower sizes because of their low efficiency and power factor. To fill this gap, the Allis-Chalmers Manufacturing Company has developed the "Synduction * motor," which is built like an induction motor but operates at synchronous speed.

Figure 21-4 shows a cross-section of a Synduction motor schematically. The stator is a smooth ring with a conventional a-c winding, just like that of a conventional induction motor. The rotor laminations contain grooves which give the rotor the appearance of having salient poles. The shaded portion of the figure is magnetic steel, whereas the light portions are nonmagnetic. A squirrel-cage winding is embedded in slots near the surface of the laminations. The rotor bars are die-cast, and the voids are also filled with die-cast metal, so that the rotor is a smooth cylinder carrying a squirrel-cage winding.

An exploded view of a Synduction motor is shown in Fig. 22-4. In outward appearance, the motor looks much like a conventional squirrel-cage induction motor. Close examination of the rotor, however, reveals the alternate bands of laminations and die-cast metal.

The peculiar shape of the rotor laminations in combination with the die-cast filled portions of the rotor have the effect of causing the reluctance of the main-axis flux path to differ greatly from that of the quadrature-axis flux path. In Fig. 21-4 the main-axis flux path

---

* Registered trade mark of the Allis-Chalmers Manufacturing Company.

SYNDUCTION MOTORS 211

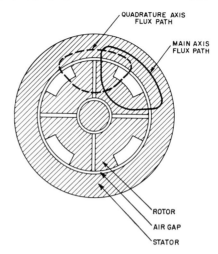

Fig. 21-4 Diagrammatic section of a Synduction motor.

is indicated by a solid line. Its reluctance is low because it includes only the normal air gap of the motor, whereas the quadrature-axis flux path has a much higher reluctance because it includes the reluctance of the die-cast filled portion of the rotor in addition to the regular air gap. Hence the quadrature-axis flux is virtually suppressed, while the main-axis flux produces an effect equivalent to that of a d-c excited salient-pole rotor.

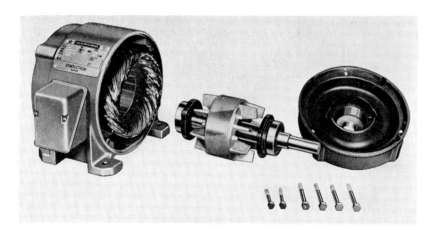

Fig. 22-4 Exploded view of a Synduction motor. (Courtesy Allis-Chalmers Manufacturing Company.)

212 CONTROL OF SYNCHRONOUS MOTORS

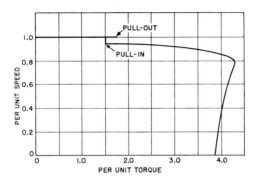

Fig. 23-4 Speed-torque curve of a Synduction motor. (Courtesy Allis-Chalmers Manufacturing Company.)

A typical speed-torque curve of a Synduction motor is shown in Fig. 23-4. Starting torque is comparatively high, higher than that of conventional squirrel-cage motors. Locked-rotor current is also comparatively high, approximately ten times full-load current.

The speed-torque curve of the Synduction motor resembles that of of a conventional squirrel-cage motor up to a speed of approximately 0.95 per unit. At this speed, the motor pulls into synchronism with a reluctance synchronizing torque of about 1.5 per unit. Once synchronized, the motor runs in synchronism at normal load, and its pull-out torque at rated voltage is of the order of 1.75 per unit. Line-voltage variations of plus or minus 10 per cent can be tolerated by the motor.

Synduction motors are started by standard squirrel-cage motor starters. Depending on load or power-system conditions, either full-voltage starting or reduced-voltage starting can be employed. Since full-load current and locked-rotor current of a Synduction motor are higher than those of a standard squirrel-cage motor of the same horsepower rating, starters are chosen one size larger than their standard a-c horsepower ratings.

BIBLIOGRAPHY

1. D. R. Shoults, C. J. Rife, and T. C. Johnson (Editor), *Electric Motors in Industry*, Chapter 6, John Wiley and Sons, New York, 1942.
2. P. B. Harwood, *Control of Industrial Motors*, Chapter 19, John Wiley and Sons, New York, 1952.
3. R. W. Jones, *Electric Control Systems*, Chapters 4, 11, John Wiley and Sons, New York, 1953.
4. D. W. McLenegan and A. G. Ferris, "Synchronous Motors, Design and Application to Meet Special Requirements," *Transactions AIEE*, June 1931.
5. M. A. Hyde, Jr., "Synchronous Motors with Phase Connected Damper Winding for High-Torque Loads," *Transactions AIEE*, June 1931.

PROBLEMS

6. C. E. Kilbourne and I. A. Terry, "Dynamic Braking of Synchronous Machines," *Transactions AIEE*, December 1932.
7. D. R. Shoults, S. B. Crary, and A. H. Lauder, "Pull-in Characteristics of Synchronous Motors," *Electrical Engineering*, December 1935. Discussion of this paper in *Electrical Engineering*, August 1936.
8. A. H. Lauder, "Salient Pole Motors out of Synchronism," *Electrical Engineering*, June 1936.
9. W. J. Boice, B. H. Caldwell, and M. N. Halberg, "Matching Synchronous Motor Excitation to a Fluctuating Load," *General Electric Review*, May 1943.
10. A. P. Burris, "Selection and Application of Synchronous Motors," *Electrical Contracting*, December 1944 and January, February, March 1945.
11. Hal Gibson, "Synchronizing Synchronous Motors," *General Electric Review*, April 1945.
12. R. R. Gobeli, "Synchronous Motors Require Multifunctional Controls," *Electrical Manufacturing*, April 1949.
13. W. A. Thomas and O. W. Whitwell, "The Selection of Field Circuit Resistance for Synchronous Motors," *Proceedings AIEE* 1951, Section T1377.
14. R. R. Gobeli, "Synchronous Motor Control . . . What Do You Expect It To Do?" *Mill and Factory*, March 1954.
15. J. Baude, "Improved Fast Acting Thermal Relay and Its Application as a Cage Winding Protective Relay for Synchronous Machines," *AIEE Paper* 55-100.
16. J. Baude, "Advancements in Synchronous Motor Control and Protection," *AIEE Paper* 55-733.
17. R. J. Dineen, "The Synduction Motor," *Allis-Chalmers Electrical Review*, 4th Quarter, 1956.
18. L. H. Harrison, "Synchronous Motors, Application, Starting, Operation," *Coal Age*, April 1957.
19. R. R. Gobeli, "Methods of Starting Synchronous Motors," *Machine Design*, February 20, 1958.
20. A. H. Hoffman, C. Raczkowski, and R. B. Squires, "Relaying for Synchronous Motor Pull-out Protection," *AIEE Paper* 59-81.
21. R. C. Moore, "Synchronous Motor Torques," *Allis-Chalmers Electrical Review*, 3rd Quarter, 1959.

PROBLEMS

1. A motor-generator set is used for supplying a source of d-c power for general use in a manufacturing plant. State some reasons why a synchronous motor may be chosen to advantage for driving the set.

2. Indicate whether the following statements are true or false, and explain your answer:
   (a) An induction motor will run at synchronous speed at no load.
   (b) Synchronous motors with no amortisseur winding cannot develop a starting torque.
   (c) Synchronous motors with amortisseur windings are started as induction motors.
   (d) To start a synchronous motor, its field must be excited.

CONTROL OF SYNCHRONOUS MOTORS

(e) The angle at which field is applied has a bearing on synchronizing torque.

(f) Field strength has no effect on the maximum torque which a synchronous motor may develop.

(g) When a synchronous motor falls out of step, its field excitation is maintained to effect self-synchronization.

(h) The power factor of a synchronous motor is shifted in a leading direction by increasing its field current.

3. Select line-contactor sizes for the following synchronous motors:
   (a) 50 hp, 220 volts, 1.0 pf
   (b) 150 hp, 440 volts, 0.8 pf
   (c) 500 hp, 2300 volts, 0.8 pf
   (d) 2000 hp, 4160 volts, 1.0 pf

4. Given a 28-pole, 440-volt, 60-cycle motor:
   (a) What is its synchronous speed?
   (b) What is its no-load speed?
   (c) What is its speed at 125 per cent load?
   (d) What is its speed at a steadily applied load of 300 per cent with normal field excitation?
   (e) What is the full-load speed if 400 volts are applied to the motor?
   (f) If the motor pull-out torque at rated voltage is 200 per cent, what will it be with a terminal voltage of 420 volts?

5. A 75-hp, 1800-rpm, 440-volt synchronous motor has a starting torque of 1.25 per unit. The motor is equipped with an autotransformer starter connected to the 65 per cent tap. Line voltage is 420 volts. What is the starting torque of the motor?

6. The motor in problem 5 has a full-load current of 86 amperes. Its locked-rotor current is 5.8 per unit. What is the starting current, as seen from the line, with a starter as specified in problem 5?

7. A 1200-rpm, 6-pole synchronous motor is able to accelerate its load as an induction motor to 1175 rpm. Field is applied by a slip-frequency relay. For what drop-out time should this relay be adjusted?

5 LOGIC FUNCTIONS AND STATIC SWITCHING

Motor controllers as discussed in the preceding chapters may be used as individual units, serving independent motor drives which function as self-contained entities without any correlation with other equipments. Or they may function in definite relation to other elements of a system, which may involve mechanical or electrical elements, or a multiplicity of motorized drives. They may form part of a continuous process system in which the various drive units are controlled automatically as a function of a prearranged program, of sensors reading process-control quantities, of limit switches indicating mechanical position, of manually actuated control-circuit devices giving overriding commands, or of a multitude of similar control signals.

Technically, all interlocking, sequencing, and other automatic control functions can be performed by magnetic control relays. However, with growing complexity of relay systems, two weaknesses become apparent. The first is the difficulty of reading complex diagrams involving a large number of coil and contact symbols. Representation by "logic functions" indicates the purpose and intended result of a control function disregarding physical appearance and detailed circuitry of the circuit elements. Logic function diagrams are more compact and represent a complex automatic sequence function in an easier to follow manner than conventional elementary diagrams.

The second problem pertains to reliability. The larger the number of contacts in an automatic control system, the greater becomes the probability of a shut down of the system, with consequent loss of income on account of loss of production due to a contact failure. Such failures can be reduced in number by proper design and choice of materials. However, since contacts are wearing parts, failures cannot be eliminated altogether. To increase the reliability of automatic control systems, static switching control devices have been developed which perform switching functions without the use of contacts.

This chapter is not intended to present a treatise on all phases of automatic control. Such treatment would greatly exceed the scope of this book. A brief introduction will be given to explain the fundamentals of logic functions and their symbolism. Likewise, this chapter will not go into details of static control unit design, but rather explain the use of these units as components in control equipments.

LOGIC CONCEPT

In an automatic control system, motors, solenoids, solenoid valves, Thrustor mechanisms, clutches, and so forth are the power converters which convert electric energy into a form of energy directly useful to the industrial process. Electrical control directs the performance of these power converters, and it also protects the equipment against malfunctioning and hazards created thereby. The conventional components by means of which control performs these functions can be divided into three major groups:

1. Control-circuit devices, such as pushbuttons, master switches, limit switches, pilot devices, sensing devices. These are the sources of information which is used to establish the requirements to be fulfilled by the control.

2. Decision-making elements, such as relays and static switching devices. These are the logic elements which make decisions based on the information.

3. Power devices, such as contactors, power-actuated switches, and the like. These are the power output elements which perform the work called for by the decision.

In logic function, electrical signals or impulses are employed in a decision-making capacity. According to Webster, logic is the science that deals with the methods of reasoning. Logic function, therefore, is simply a means of expressing predetermined reasoning with electrical signals. The logic-function concept to accomplish control performance makes use of some of the very same human reasoning processes that are applied in everyday life. Five functions are needed to express control performance in logic terminology, namely:

AND
OR
NOT
DELAY
MEMORY

AND, OR, NOT FUNCTIONS

Although logic functions need not be exact duplicates of relays, relay circuits can be expressed by logic functions. Therefore, basic logic functions and their symbols will be explained by equivalent basic relay circuits. Symbols for logic elements have been standardized by NEMA, and complex diagrams are made up of a comparatively few basic symbols.

Logic element symbols describe the status of an output as a function of the condition of one or several inputs. In the equivalent basic relay circuit, the output is the circuit established by a relay contact. The inputs are contacts which cause the relay coil to be energized or deenergized. The logic element symbol is not specific in regard to the character of the input and output. While they may represent contact position, it is better to think of input and output as a condition of flow of electric energy. Most commonly, input and output are expressed as the presence or absence of a voltage which represents a control signal.

Logic element symbols and equivalent basic relay circuits for *AND*, *OR*, and *NOT* functions are given in Fig. 1-5. These functions have

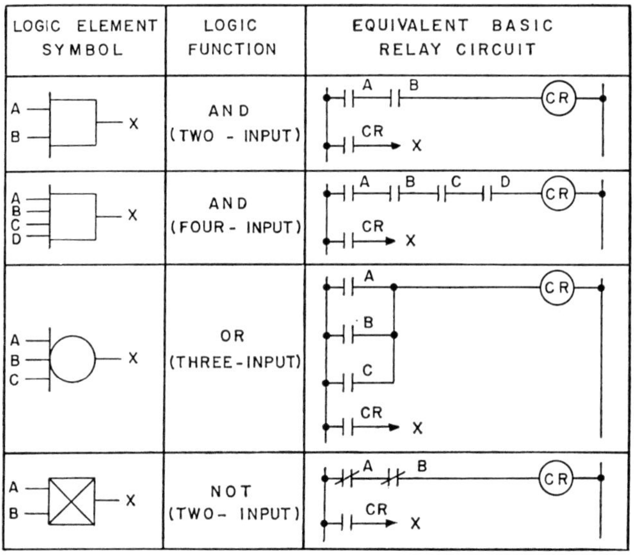

LOGIC ELEMENT SYMBOL	LOGIC FUNCTION	EQUIVALENT BASIC RELAY CIRCUIT
A ──⊐── x B ──	AND (TWO - INPUT)	
A ──⊐── x B ── C ── D ──	AND (FOUR - INPUT)	
A ──⊐── x B ── C ──	OR (THREE - INPUT)	
A ──⊠── x B ──	NOT (TWO - INPUT)	

Fig. 1-5 Logic element symbols and equivalent basic relay circuits for *AND*, *OR*, and *NOT* functions.

the common feature that the output follows instantly any changes in the input or inputs. The most common pattern for logic elements is that several inputs result in a single output. In relays, a multiplicity of outputs can be obtained by using multicontact relays. This feature is an important one to remember when laying out circuits with logic elements.

The AND function describes an element which produces an output only when every input is energized. For example, a two-input element has two inputs, A and B, and one output, X. In equivalent relay circuitry, relay CR is energized only if contacts A and B are closed. A and B may be contacts on two other relays, two limit-switch contacts indicating that two positional requirements are fulfilled, two interlocks on starters permitting the sequencing of another starter controlled by CR, just to mention a few possible interpretations. Output X disappears as soon as either input A or input B is deenergized. In equivalent relay parlance, CR drops out as soon as contact A or contact B opens. Similarly a four-input AND element produces an output X if all four inputs A, B, C, D are energized. Its output is deenergized when any one of its four inputs is deenergized.

The OR function describes an element having a multiplicity of inputs which produces an output if one or more of its inputs are energized. In equivalent relay circuitry, the OR function is represented by a relay, the coil of which is energized when any one of several parallel contacts is closed. An output X is obtained if input A or input B or input C or any combination of inputs, is energized. Output X disappears when all inputs disappear.

The NOT function describes an element with one or several inputs which produces an output only if its input is *not* energized. In conventional relay circuitry, the NOT function represents the case that a relay is energized through the normally closed contacts of other devices, indicating that these devices are not energized. In logic circuitry, NOT elements are used to advantage for obtaining inverse functions, off signals and the like.

DELAY FUNCTION

Machine specifications often require the introduction of a time delay before a desired operation occurs. The $DELAY$ function indicates that ultimately the output will correspond to the condition of the input. However, a time interval may elapse before correspondence between

DELAY FUNCTION

LOGIC ELEMENT SYMBOL	LOGIC FUNCTION	EQUIVALENT BASIC RELAY CIRCUIT
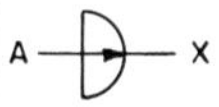 A —⟩— X	TIME DELAY ENERGIZING	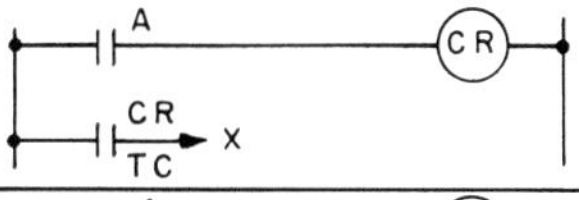 A ──┤├──────────(CR)── ──┤├─→ X $\;\;\;$ CR/TC
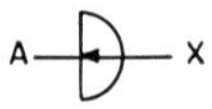 A —⟨— X	TIME DELAY DEENERGIZING	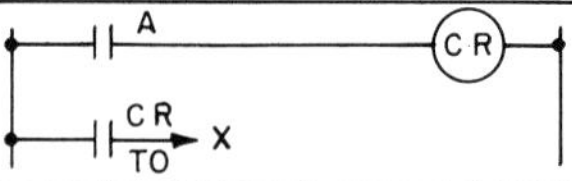 A ──┤├──────────(CR)── ──┤├─→ X $\;\;\;$ CR/TO
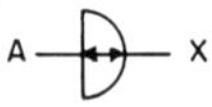 A —⟨⟩— X	TIME DELAY ENERGIZING AND DEENERGIZING	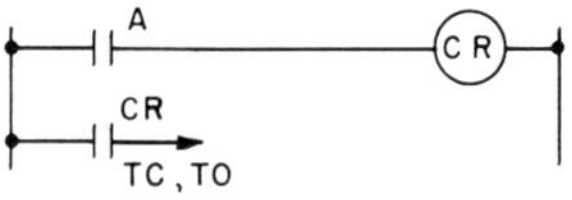 A ──┤├──────────(CR)── ──┤├─→ $\;\;\;$ CR/TC,TO
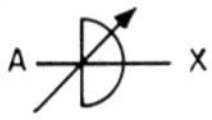 A —↗⟨— X	ADJUSTABLE TIME DELAY	MAY BE APPLIED TO ANY OF ABOVE CIRCUITS

Fig. 2-5 Logic element symbols and equivalent basic relay circuits for $DELAY$ functions.

output and input is attained. Figure 2-5 shows various $DELAY$ function symbols. They all contain a half-circle as a general delay symbol, the arrows indicating the direction in which the delay functions.

$DELAY$ energizing means that, when input A is energized, output X is energized with a time delay. When input A is deenergized, output X disappears instantly. It is the equivalent of a time relay with a time closing (TC) and instantaneously opening contact.

$DELAY$ deenergizing means that, when input A is energized, output X is energized instantly. When input A is deenergized, output X disappears after a time delay. It is the equivalent of a time relay with an instantaneously closing, time opening (TO) contact.

$DELAY$ energizing and deenergizing means that, when input A is energized, output X is energized with a time delay. When input A is deenergized, output X disappears after a time delay. It is the equivalent of a time relay with a time closing (TC) and time opening (TO) contact.

The symbol for adjustable time delay may be superimposed on the foregoing $DELAY$ function symbols. It denotes a property of the logic element, namely that the time-delay interval may be adjusted. The direction of the time delay is indicated solely by the horizontal arrows.

220   LOGIC FUNCTIONS AND STATIC SWITCHING

MEMORY FUNCTION

A control input may be momentary; that is, a control action is initiated and established, but the initiating input then disappears. But it may be desired to continue the control action, as with momentary-contact pushbuttons, spring-return master switches, and limit switches that close only at definite points of machine travel. This control function is accomplished by MEMORY elements. They have two inputs. One input energizes the output, which is maintained after the first input ceases. A second input is required to deenergize the output. There may be two outputs, one corresponding to the energized, and one to the deenergized condition of the MEMORY elements.

Figure 3-5 indicates logic symbols and equivalent basic relay circuits for several types of MEMORY units. The convention in arranging input signals is such that the upper input A is the one which causes output X to be energized. Lower input B is the one which causes output

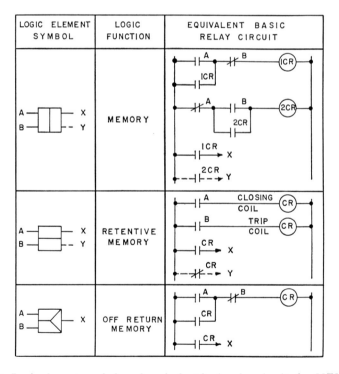

Fig. 3-5   Logic element symbols and equivalent basic relay circuits for MEMORY functions.

MEMORY FUNCTION 221

X to be deenergized. There may be a second output Y (indicated by broken lines) which is the inverse function of X; that is, Y is energized when X is deenergized, and vice versa.

A simple *MEMORY* element retains the condition of output corresponding to the input last energized, as long as control-circuit power is maintained. This means that a simple *MEMORY* element ceases to retain its memory function upon an interruption of control power, and that, upon return of power, an input signal is required to set up the desired output by energizing one of the inputs. One way of obtaining a corresponding performance from a relay circuit would be to use two control relays, only one of which could be energized at one time, and each relay being associated with a distinct function. For example, let $1CR$ be the slow-speed relay and $2CR$ the fast-speed relay of a two-speed induction motor. Let inputs A and B be the momentary-contact (including back contacts) slow-speed and fast-speed pushbuttons. Pressing pushbutton A momentarily energizes relay $1CR$, which seals itself in and produces output X. When pushbutton B is pressed momentarily, relay $1CR$ is dropped out, and relay $2CR$ is energized and seals itself in. Output X is canceled, and output Y is energized. When control power is interrupted, $1CR$ and $2CR$ drop out, and outputs X and Y disappear. When control power is reestablished, one of the inputs has to be energized (that is, one of the pushbuttons has to be pressed) to obtain an output (either slow- or fast-speed operation).

A *RETENTIVE MEMORY* element retains the condition of output corresponding to the input last energized, irrespective of whether control-circuit power is maintained or not. A simple relay equivalent is the mechanically held (or latched-in) relay. Momentarily closing contact A energizes the closing coil of relay CR, which is held in its closed position by a mechanical latch. Relay position is maintained when control power is interrupted, since a separate power input is required to drop out the relay. Momentarily closing contact B energizes the trip coil of relay CR, which breaks the mechanical latch and causes the relay armature to move in the dropped-out position. Outputs X and Y are represented by normally open and normally closed contacts of relay CR.

An *OFF RETURN MEMORY* element retains the condition of output corresponding to the input last energized, but returns to the off position upon interruption of control power. It resembles a simple *MEMORY* element in that it does not maintain its output in case of

222 LOGIC FUNCTIONS AND STATIC SWITCHING

power interruption. It differs in that the element resets itself to a definite output condition. A simple relay equivalent would be a start-stop pushbutton circuit. Let A be a momentary-contact start button energizing relay CR, which seals itself in. Relay CR is dropped out (or returned to the off position) by momentarily pressing stop button B. The same return to the off position occurs when control power is interrupted. In order to energize output X again, it is necessary to energize input A. Thus, the *OFF RETURN MEMORY* element provides undervoltage protection inherently.

AUXILIARY FUNCTIONS

Any control relaying diagram can be drawn using only the logic symbols described in the preceding paragraphs. However, it is convenient to introduce a few auxiliary functions which permit representing the power output and certain peculiar features of static switching elements in a fashion that is compatible with the logic function representation of the control circuitry. Symbols for these auxiliary functions are shown in Fig. 4-5.

Relays as well as static switching devices do not have outputs (or contact capacities) of a sufficiently high power level to actuate the power converters which produce the output of the industrial processes. Amplifiers are used to step up the power level. An amplifier can be described as a device in which an input signal controls a local source of power to produce an output enlarged relative to the input. The

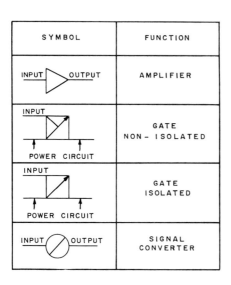

Fig. 4-5 Auxiliary function symbols for use in logic function diagrams.

COMBINING LOGIC ELEMENTS

amplifier symbol is a triangle, the same as that used conventionally in electronic and communications diagrams and as recognized in American Standard Y32.2. Commonly used amplifiers in relay function systems are magnetic contactors. In logic element systems, amplifiers produce sufficient output to actuate these contactors.

A gate is a control element often found in static switching units which inhibits or permits current flow in a power circuit. More specifically, in logic elements a gate permits current flow when the control input is energized. The terms "nonisolated" and "isolated" denote whether power and control circuits are tied together by a common connection or isolated from each other. Inherently, a gate may act as a power amplifier, a very small control input permitting the flow of a larger amount of energy in the power circuit. The classical example of a gate is the grid of an electron tube triode. Other examples of gates are control electrodes of transistors and controlled semiconductor rectifiers, or control windings of magnetic amplifiers.

In control relay circuitry, inputs and outputs are the closing or opening of contacts which provide or close a path for the flow of control-power currents. In static switching devices, input and output signals often appear as some other form of electric energy. Most often they are voltages. For reasons of mechanical sturdiness and economy, conventional contact-making control-circuit devices are often used to provide important information about machine performance, as well as to override manual signals. Most frequently used control-circuit devices are limit switches and pushbuttons. The input signals provided by the contact closing of these devices cannot, in general, be accepted directly as input signals to static switching elements. Signal converters have to be employed for converting a pilot signal into a logic input compatible with the static switching units. In most cases, signal converters are voltage-divider networks which, when a pilot device contact closes or opens, cause an input voltage to be applied to the static switching unit.

COMBINING LOGIC ELEMENTS

Initial understanding of logic elements and their circuitry can best be gained by comparing logic circuitry with corresponding relay circuitry. However, once the fundamentals are understood, one should strive to identify the required control functions in terms of machine functions and to develop a logic diagram without giving any consideration to the manner in which the problem would be solved by the use of magnetic relays. It is natural that one accustomed to

224   LOGIC FUNCTIONS AND STATIC SWITCHING

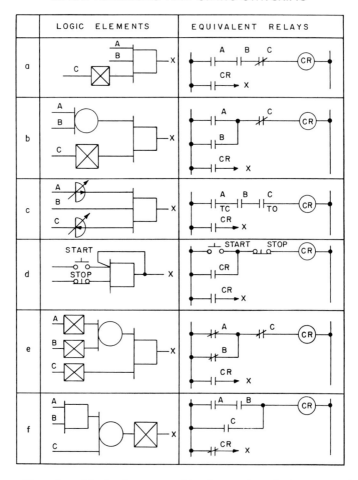

Fig. 5-5 Examples of logic element combinations and equivalent relay circuits.

magnetic control would first lay out a conventional elementary diagram and translate it into logic symbols. However, the preferred method is to start out directly with a logic diagram and work out static switching unit connections from the logic circuits.

Figure 5-5 illustrates how individual logic elements are combined into more complex circuit elements. The left column shows logic circuits, whereas the right column shows relay circuits obtaining the same output performance.

Sketch *a* is a combination of *AND* and *NOT* functions. An output

X is obtained if inputs A and B are energized *and* input C is *not* energized.

Sketch b contains AND, OR, and NOT functions. Output X is energized when input A or B is energized *and* input C is *not* energized.

Sketch c illustrates the introduction of time delay. Output X is energized when inputs A and B and C are energized. Input A includes a time delay upon closing, and input C includes a time delay upon opening. Both time delays are adjustable.

Sketch d indicates how an AND unit can be utilized for a simple start-stop circuit. Output X is energized when both the start *and* the stop button are closed. The output is connected to the start-button input as a sealback, so that the output is maintained when the start button is released. The output is deenergized when the stop button is pressed or when control power is lost, which is equivalent to deenergizing the stop-button input. This circuit combination is functionally equivalent to the OFF $RETURN$ $MEMORY$ element.

In logic circuitry, it is sometimes advantageous to invert the output because available circuit components may limit the flexibility of circuit design. The circuits shown in sketches e and f are equivalent. Consider first sketch e. Output X is energized when inputs A or B are *not* energized *and* when input C is *not* energized. Expressed differently, output X is *not* energized when inputs A and B, *or* input C, or all three inputs are energized. This is exactly the condition indicated by sketch f. It is noted that, when the output is inverted, the AND and OR functions governing the relationship of inputs have to be interchanged.

POINTERS ON CIRCUIT DESIGN

In starting to work with logic elements, one who is accustomed to laying out conventional magnetic control circuits will be tempted first to design the control circuits for magnetic devices and then to translate the magnetic diagram into a logic diagram. Such a procedure entails waste of time and effort, which can be avoided if the control engineer designs the circuits directly by logic functions. Circuit details vary considerably from job to job, and no universal theory can be advanced on how to approach the circuit design for a specific application. Inspection and common sense are the most important tools to be used by the circuit designer. Advanced circuit theory, Boolean algebra, and similar tools offer advantages to those who are engaged in computer work and very complex control systems. These tools are generally not needed for the solution of industrial problems, and hence are deemed to be beyond the scope of this book.

226 LOGIC FUNCTIONS AND STATIC SWITCHING

The following paragraphs contain some pointers which are offered as suggestions on how to approach the design of a logic function diagram for a specific application. Design of control circuits is greatly simplified if these steps are followed:

1. Study machine operation and write down each functional step of machine performance.
2. Make a sequence analysis, and write down the function and operation of each auxiliary control-circuit device used on the machine.
3. Draw a logic diagram of each machine function by translating the word picture into graphical symbols, and combine the various functions into a composite diagram.
4. Inspect the diagram, minimize circuitry, and save devices by simplifying circuits, and adapt circuit details to limitations of available logic elements.

It is axiomatic that the machine operation must be studied out functionally before the control circuit can be designed. Writing down each function is of great help in making certain that no detail is overlooked.

In analyzing the required control sequence, a list should be prepared of all control-circuit devices used in controlling the operation of the machine. Names should be assigned to the devices in accordance with the function with which they are associated. Although there are some contactless control-circuit devices on the market which may be connected directly to logic elements, most devices performing pilot functions are still conventional contact devices, and it is advisable to write down the number and operation of the contacts available on each device. It is also advisable to write down whether contacts are maintained or momentary. An example of a control sequence chart is this:

Device	Contacts	Device Action	Contact Action	Function
Clamps retracted limit switch	1—NO 1—NC	Spring return to neutral	Maintained	Interlocks with advance transfer
Start pushbutton	1—NO	Spring return to open	Momentary	Starts cycle

POINTERS ON CIRCUIT DESIGN 227

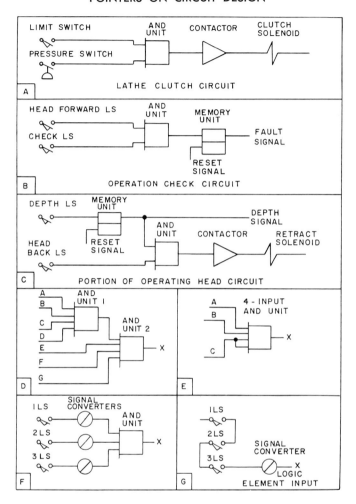

Fig. 6-5 Examples of logic circuits utilizing *AND* units.

The next step is to design the logic circuits. Working on the machine as a whole is cumbersome and confusing; it is much better to lay out circuits for individual machine functions first and then to combine them. To illustrate the manner in which circuits are laid out, various examples of machine circuits, involving *AND* logic elements, are illustrated in Fig. 6-5.

Circuit *A* is the final stage of the clutch circuit of a lathe. The limit switch indicates work in position. When it *and* the air pressure switch are closed, indicating air pressure available, the clutch solenoid is energized. The AND unit energizes an amplifier, which may be a contactor, which in turn actuates the solenoid coil. When air pressure is lost and the pressure switch opens, the clutch solenoid is deenergized.

A check circuit B supervises a machine operation and assures that the operating head is in proper sequence. Normally the operating head should have left the forward position, before the check limit switch closes. If the head is still forward, the closing of the head-forward limit switch *and* the check limit switch energizes a $MEMORY$ unit, and its output fault signal is used to stop the machine and possibly give a signal. After the cause of the fault has been removed (head-forward and/or check limit switches open) the $MEMORY$ unit is reset, the fault signal is deenergized, and normal operation is resumed.

A portion of the circuit of an operating head is given in C. As the head moves forward, the head-back limit switch closes. When the head reaches the forward position, it registers depth by closing the depth limit switch. With the depth limit switch *and* the head-back limit switch closed, the retract solenoid is energized and retract motion is initiated. Since the depth limit switch opens as soon as the head leaves the register position, a $MEMORY$ unit is inserted to maintain the retract signal. When the head is retracted, the head-back limit switch opens, the retract solenoid is deenergized, and the head is ready for another forward stroke. The $MEMORY$ unit is reset by a suitable reset signal, for instance, when the head is started forward.

Logic elements are designed for a certain number of inputs, just as relays are designed for a specific number of contacts. Circuit D illustrates how to combine several AND units if there are more inputs than can be accommodated on a single unit. The example shows how to serve seven inputs with two four-input AND units. Conversely, to obtain an output, all inputs of the AND unit must be energized. Circuit E indicates how two inputs of an AND unit are paralleled so that a four-input AND unit can be used in conjunction with three inputs.

In the preceding circuits, pilot-device inputs are shown feeding directly into the logic units. In reality, signal converters are used to feed contact-type inputs into logic elements. In order to save logic elements and signal converters, resulting in a simplified control circuit, contact-type pilot circuits should be combined as far as possible before an input is fed into logic elements. Circuit F shows what would happen if three limit-switch inputs, each with its own signal converter, were fed into an AND unit. Connecting the three limit switches in

POINTERS ON CIRCUIT DESIGN 229

series and to a signal converter, as indicated in circuit G, obtains the same logic signal at a saving of two signal converters and one AND unit.

Examples of logic circuits utilizing OR and NOT elements are indicated in Fig. 7-5. Circuit A illustrates the control of a loading solenoid, actuating a loading mechanism. Automatic operation results when the loading arm is up (as indicated by loading-arm-up

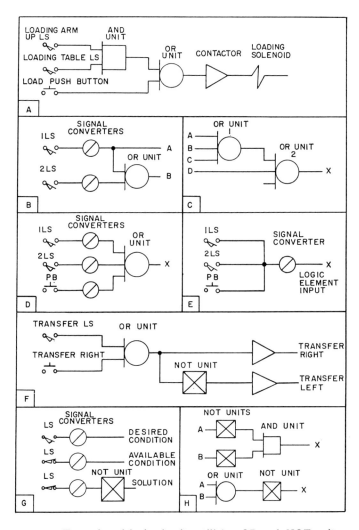

Fig. 7-5 Examples of logic circuits utilizing OR and NOT units.

limit switch being closed) *and* the loading table is in proper position (as indicated by the loading-table limit switch being closed), *or* when the loading solenoid is actuated manually by means of a pushbutton. *OR* elements are inherently isolating elements which may be used to isolate signals. In circuit B, limit switch $1LS$ energizes outputs A and B, whereas limit switch $2LS$ energizes only output B.

Several *OR* units may be combined if more inputs are to be accommodated than can be handled by a single unit. Circuit C illustrates how two three-input *OR* units are combined to respond to four inputs. Since any one of several inputs energizes the output of an *OR* unit, inputs on elements, which are not utilized, are simply left unconnected. This condition is illustrated by *OR* unit 2.

A saving in logic elements can be effected if control-circuit devices are interconnected directly before their combined input is fed into a logic element. In circuit D, two limit switches and one pushbutton are connected individually to an *OR* unit. The same results are obtained with circuit E at a saving of an *OR* unit and two signal converters.

Circuit F illustrates the use of a *NOT* unit for obtaining an inverted signal. When a transfer limit switch *or* a transfer-right pushbutton are closed, a machine carriage is transferred to the right. When both the limit switch and the pushbutton are open, the *OR* unit produces no output, and the *NOT* unit produces an output causing transfer to the left.

NOT units can be used to advantage to overcome restrictions on available control-circuit contacts on the machine. An example is illustrated in G. The desired condition is to obtain an output when a limit switch closes its contact while a machine part is in a certain position. Such a contact, however, is not available. Instead, a limit-switch contact is available which is normally closed and opens when the machine position is reached. Combining this limit switch with a *NOT* unit obtains the desired output condition.

Inversion may be obtained by the use of either *AND* units or *OR* units. In H, two circuits are shown which produce identical results. Two inputs, A and B, are to produce an inverted output X. This result can be obtained by the use of one *AND* unit and two *NOT* units, or by using one *OR* unit and one *NOT* unit.

DELAY units are versatile units which may be used not only for obtaining a straight time delay in energizing or deenergizing a logic output, but they may also be combined with other logic elements to produce control outputs of a limited time duration. Figure 8-5 gives examples of circuits utilizing *DELAY* units.

PLANER CONTROL

Circuit A obtains a straight time delay in energizing logic output X after the closing of the limit switch. In the sketches below the circuit, the cross-hatched areas indicate the time during which the limit switch is closed, and during which the output signal X is energized. When the limit switch opens, X is deenergized instantly. Time delay is adjustable. By using $DELAY$ units with time delay on deenergizing, or delay in both directions of operation, a delay in deenergizing, or in both directions, of output X may be obtained. With such circuits, the output signal is lost in case of loss of power.

Combining $DELAY$ and $MEMORY$ functions, output can be maintained in case of loss of power. In circuit B, closing of the limit switch immediately energizes output X, and this output remains energized upon loss of power. When the limit switch opens, the NOT unit produces an output which, through the $DELAY$ unit, energizes the "off" part of the $MEMORY$ unit to deenergize output X. Hence, after the limit switch opens, output X remains energized for an adjustable period of time.

In circuit C, with the limit switch open, one input to the AND unit is energized because the NOT unit produces an output. When the limit switch closes, the second input to the AND unit is energized, and an output X results. At the same time, the $DELAY$ unit receives an input and, after a time delay, its output energizes the NOT unit resulting in a loss of one of the AND inputs. X is deenergized. Thus circuit C obtains an output signal lasting an adjustable period of time after the limit switch has been closed.

An opposite operation results from circuit D. With the limit switch open, the AND unit receives one input through NOT unit 1; hence there is no input to NOT unit 2 and output X is energized. When the limit switch closes, both inputs to the AND unit are energized; hence NOT unit 2 receives an input and output X is deenergized. After a time delay, NOT unit 1 receives an input, thus deenergizing one input to the AND unit, which deenergizes its output. Since NOT unit 2 no longer receives an input, output X is reenergized. Circuit D obtains the interruption of an output signal for an adjustable period of time after the limit switch has closed.

PLANER CONTROL

An example will now be given of a complete logic circuit for a production machine. The example chosen is the reversing drive of a planer table. The arrangement of power- and control-circuit devices

232 LOGIC FUNCTIONS AND STATIC SWITCHING

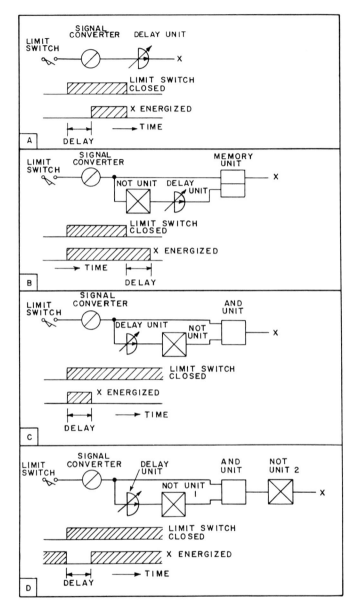

Fig. 8-5 Examples of logic circuits utilizing *DELAY* units.

PLANER CONTROL 233

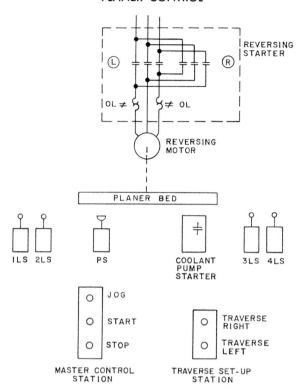

Fig. 9-5 Arrangement of power- and control-circuit devices of planer drive.

needed for the control of the planer is indicated schematically in Fig. 9-5. For the sake of simplicity, a squirrel-cage induction motor is assumed to be the driving motor. In its place an adjustable-speed a-c or d-c motor may be used. In that case, speed control would be added to the reversing control, without any basic changes in the reversing logic circuitry.

The following devices are required for the automatic planer reversing control:

1 reversing starter containing left (L) traverse and right (R) traverse contactors and overload (OL) relays.
1 starter for coolant pump, manually or magnetically actuated, with a control-circuit interlock.
2 limit switches ($1LS$, $4LS$) actuated by the planer bed, to act as safety stops at extreme ends of left and right travel.

2 limit switches (2*LS*, 3*LS*) actuated by the planer bed, adjustable for position of actuation, to effect automatic reversal of planer bed at the ends of the desired left and right traverse travel, as determined by work piece size.

1 pressure switch (*PS*) to indicate the presence of lubricating-oil pressure.

1 master control station containing jog, start, and stop pushbuttons.

1 traverse setup station containing pushbuttons for right and left traverse.

Functional specifications for the control of the planer motor are as follows:

1. The motor is to be started manually by a start pushbutton, provided coolant-pump motor runs and lubricating-oil pressure is available. Overtravel limit switches (1*LS*, 4*LS*) stop motor at extreme ends of right and left traverse travel. Overload protection is by overload relays *OL*, which trip to stop the motor.

2. The stop pushbutton is used for manual stopping of the motor. Undervoltage protection is to be provided. Jogging shall be possible at any time by means of the jog pushbutton.

3. Once the traverse motor is started, it is reversed automatically at the end of the right and left stroke, under the control of limit switches 2*LS* and 3*LS*. Right and left traverse pushbuttons permit manual selection of initial direction of travel when the planer motor is started.

The complete logic diagram of the reversing planer drive is given in Fig. 10-5. There are two major sections to the circuits. One is the control of the starting and stopping of the motor, and the other is the automatic reversing control.

Reading the diagram from the top, *AND* unit 1 has two inputs: one is provided by the start button, the other by the stop button in series with the overtravel limit switches. The sealback circuit of the *AND* unit obtains undervoltage protection in the same manner as start-stop pushbutton magnetic control. Pressing the stop button, or tripping of one overtravel limit switch deenergizes the *AND* input and results in the loss of the *AND* unit 1 output. This output is fed into an *OR* unit, the other input being provided by the jog pushbutton. The *OR* unit provides an output when either the regular start circuit is energized or when it is bypassed by the jog button. This means that jogging is possible at any time, either for tool or work adjustment, or for backing the planer table out of the overtravel limits.

PLANER CONTROL

Control-Sequence Chart

Device	Contacts	Device Action	Contact Action	Function
Start pushbutton	1—NO	Spring return to open	Momentary	Starts motor
Stop pushbutton	1—NC	Spring return to close	Momentary	Stops motor
Jog pushbutton	1—NO	Spring return to open	Momentary	Jogs motor
Right traverse pushbutton	1—NO	Spring return to open	Momentary	Sets up right traverse motion
Left traverse pushbutton	1—NO	Spring return to open	Momentary	Sets up left traverse motion
Limit switch $1LS$	1—NC	Spring return to neutral	Maintained	Stops motor at extreme left travel
Limit switch $2LS$	1—NO	Spring return to neutral	Maintained	Reverses motor at end of left stroke
Limit switch $3LS$	1—NO	Spring return to neutral	Maintained	Reverses motor at end of right stroke
Limit switch $4LS$	1—NC	Spring return to neutral	Maintained	Stops motor at extreme right travel
Pressure switch PS	1—NO	Closes on rising pressure	Maintained	Stops motor on loss of lubricating-oil pressure

The output of the OR unit is one of the inputs of AND unit 2, which serves to prevent operation of the traverse motion in case of an unsafe operating condition. AND unit 2 has four inputs, the remaining three being energized by pressure switch PS, an interlock on the coolant-pump starter, and the contacts of overload relays OL. Since all four inputs must be energized to obtain an output from AND unit 2, the traverse motor is prevented from starting (or stopped in case it had

236 LOGIC FUNCTIONS AND STATIC SWITCHING

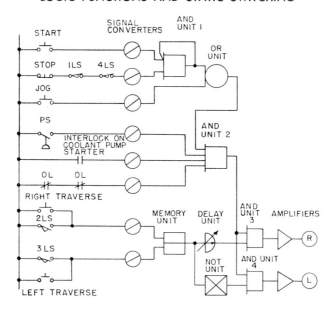

Fig. 10-5 Logic diagram of automatic reversing control of planer drive.

been running), when lubricating-oil pressure fails (which would cause damage to the table guides), the coolant pump stops (which would cause overheating and damage to the tools and the work piece), or the motor has been overloaded. The output of AND unit 2 energizes one of the inputs of AND units 3 and 4, which form part of the automatic reversing circuit.

Selection of right (contactor R energized) or left (contactor L energized) traverse direction depends on the condition of the $MEMORY$ unit. When the equipment is initially started, the desired direction of travel may be selected by one of the traverse pushbuttons. Pressing the right traverse pushbutton turns the $MEMORY$ unit "on." It produces an output which energizes one of the inputs of AND unit 3. A $DELAY$ unit introduces a time delay for the following reason: right traverse is assumed to be the cutting portion of the cycle, and upon automatic reversing the time delay assures that the tool has assumed its proper position before a cut is started. If the start circuit has been energized and AND unit 2 produces an output, AND unit 3 energizes contactor R, and the motor traverses right.

At the end of the intended stroke, limit switch $3LS$ closes momentarily and turns the $MEMORY$ unit "off." Its output disappears;

hence *AND* unit 3 has no output, and contactor *R* drops out. Since the *MEMORY* unit no longer has an output, the *NOT* unit produces an output which is fed into *AND* unit 4. Its output energizes contactor *L*, and the motor traverses left. This being the return stroke, no time delay is introduced. At the end of the left stroke, limit switch *2LS* closes momentarily and the *MEMORY* unit is reset. Contactor *L* is dropped out, and contactor *R* is energized. This cycle is repeated until one of the inputs of *AND* unit 2 is deenergized, causing the drive to shut down. Jogging is possible at any time in either direction; the desired direction can be selected by pressing the right or left traverse pushbutton, irrespective of the direction for which the *MEMORY* unit was set up last.

PRINCIPLES OF STATIC SWITCHING

Static switching has not been introduced to industry for the purpose of reducing the initial cost of control equipment. At the present state of the art, magnetic relays are less expensive. Likewise, static switching does not reduce control panel size. With progress in the art of static switching, these conditions may change and static switching may, in the future, become more attractive from the point of view of cost and space.

Presently, the inducements for application of static switching are limitations on contact reliability and mechanical life of magnetic devices. The more complex control systems become, the greater is the probability of contact failure. Increasing the speed of manufacturing operations and processes increases the demands on device life. Many thousands of dollars may be lost per hour when an automated manufacturing process is shut down. Industrial-control-type relays having a mechanical life of 10,000,000 operations are rather common. A life of 20,000,000 operations can be realized with modern designs. However, with presently known magnetic and conducting materials, there is no prospect that this life figure can be substantially increased by further refinements in design. To increase device life, mechanical and electrical wear has to be eliminated. Today, industry demands a device life of 50,000,000 operations, which means that relays have to be replaced by nonwearing switching devices.

To explain the principles of static switching, it is helpful to visualize the performance of a conventional magnetic relay as a logic element, as indicated in Fig. 11-5. Input to a relay is the energy absorbed by the relay coil. It can be measured by the magnitude of the voltage applied to the coil terminals. Relay output is the condition of its con-

238 LOGIC FUNCTIONS AND STATIC SWITCHING

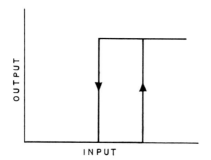

Fig. 11-5 Performance of a magnetic relay as a logic element.

tacts. A relay has two stable operating conditions: it is either picked up or dropped out. Thus, it is a "bistable" device. When the coil voltage is zero, relay output is zero. Increasing the coil voltage does not change the relay output until the pick-up voltage is reached, at which point the relay output jumps to its "full on" value. Increasing the coil voltage beyond the pick-up value produces no change in relay output. Reducing the coil voltage does not change the relay output until the drop-out voltage is reached, at which point relay output suddenly drops to zero. Since drop-out voltage is always lower than pick-up voltage, the relay output characteristic forms a "loop."

Devices presently used for static (or contactless) switching are magnetic amplifiers and transistors. The technology of these devices is moving rapidly, and considerable development work is going on, particularly in the area of semiconductor devices. Many changes in the design of static switching devices can be expected in the near future. These changes, however, will not affect the application and utilization of such devices as logic function elements. Therefore this chapter will not go into details describing specific designs and internal circuitry, and the student interested in such detail should consult papers and articles on these subjects, as well as literature available from the manufacturers.

Magnetic amplifiers utilize the change in impedance of a coil arranged on a magnetic core, because of saturation of the magnetic core. The state of saturation of the magnetic core can be influenced by a winding excited with adjustable direct current.

A magnetic amplifier utilizes the saturable reactor principle, so that, with proper design of magnetic and electric circuits, a small amount of input power can control a much larger output power. Figure 12-5A indicates schematically the basic components of a magnetic amplifier. The magnetic core carries two windings. The d-c winding is the

PRINCIPLES OF STATIC SWITCHING 239

control winding, variable d-c current changing the state of saturation of the core. The gate winding is connected to an a-c power source through a half-wave rectifier. Its impedance changes with the saturation of the core, and it thus controls the flow of current through the load just as a gate controls the flow of traffic.

This magnetic amplifier is so proportioned that, with no d-c control current flowing, the core is fully saturated by the load current; hence this amplifier is called "self-saturating." By exciting the d-c control winding in a sense to reduce core saturation, the impedance of the gate

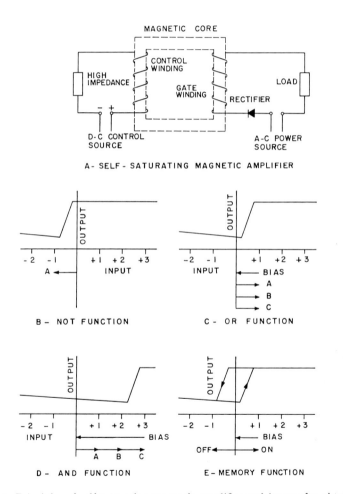

Fig. 12-5 Principles of self-saturating magnetic amplifier and its use for obtaining logic switching functions.

winding is increased, and load-current flow is throttled. Magnetic amplifiers are inherently "analog" devices; that is, their output varies as a function of input. By proper design and choice of control input, magnetic amplifiers can be used as "digital" or "bistable" devices; that is, their output is either turned on or turned off. In addition to the control winding indicated in Fig. 12-5A, other control windings may be arranged on the core, which will superimpose additional control inputs and which may be used for introducing a bias or a feedback.

Special ferro-magnetic materials are available having a sharply defined saturation point, beyond which the reluctance of the core material does not change. Such materials also may have very narrow hysteresis loops, so that the effect of hysteresis can be neglected in explaining the characteristic of the magnetic amplifier, and its transfer characteristic may be idealized and shown as a line. The natural transfer characteristic of the self-saturating magnetic amplifier is gven in Fig. 12-5B. With no d-c control current, output or load current through the gate winding is full on. The input scale is marked in arbitrary units. Applying d-c control input in a positive direction has no effect on output, since the state of saturation of the core is not changed. Applying a d-c control input A in a negative direction turns off the output, since the core is unsaturated and the impedance of the gate winding becomes maximum. Thus a self-saturating magnetic amplifier can be used as a *NOT* function device because, with no input, there is output. With input applied in the proper sense, output is de-energized.

Applying a bias through a separate bias winding, the characteristic can be shifted so that, with no control input, the output is turned off. Introducing input A *or* input B *or* input C of proper magnitude and direction, or any combination of inputs, overcomes the effect of the bias and turns on the output. This condition is illustrated in Fig. 12-5C. Such a device performs an *OR* function. Introducing a greater amount of bias, as indicated in Fig. 12-5D, the output is turned off, with zero control input. Introducing input A *and* input B *and* input C overcomes the effect of the bias and turns on the output. This combination performs an *AND* function. *DELAY* functions are obtained by internal circuitry and/or the use of additional cores, so as to obtain a time interval between the application of the control input signal and its effect on the magnetic status of the core carrying the gate winding.

Introducing a properly designed feedback, a loop can be placed in the magnetic amplifier transfer characteristic. This means that once the output has been established, a significant change in input is re-

PRINCIPLES OF STATIC SWITCHING 241

quired to turn off the output, as indicated in Fig. 12-5E. Introducing a bias, the loop can be so positioned that, with zero input, the output can be either on or off. Such a unit can be used to perform a *MEMORY* function. Applying a positive input turns the output on, and it stays on when the input is removed. Applying a negative input causes the output to be turned off, and it stays turned off when the input is removed.

Transistors can be employed in amplifiers having characteristics similar to those of magnetic amplifiers. A basic transistor amplifier circuit and its somewhat idealized transfer characteristics are indicated in Fig. 13-5. Varying the input voltage results in a change in output current, hence in output voltage across the load resistor. Output rises from zero to its full-on value over a comparatively limited range in input. Further increase in input results in no change in output. By applying inputs in units of sufficient magnitude to drive the amplifier through its full range, the amplifier can be used as a bistable device to perform static switching operations. By proper internal circuitry and the application of appropriate bias and feedback signals, any of the basic logic functions can be performed.

Several static switching systems are commercially available. Some

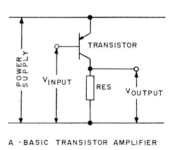

A - BASIC TRANSISTOR AMPLIFIER

Fig. 13-5 Principles of transistor amplifier and basic transfer characteristic.

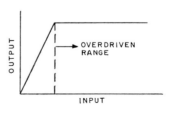

B - AMPLIFIER TRANSFER CHARACTERISTIC

use magnetic units; others use transistors. Operating experience in industry will eventually settle the relative merits of the two basic units. When static switching units are assembled and packaged into a complete switching system, the following points should be given proper consideration.

1. Each static switching logic element contains a number of small components. Each logic unit should be so arranged as a module that it can be easily removed and replaced in case of trouble. Units should be repaired in a shop rather than on the panel. Plug-in construction is preferred.

2. Because of the complexity of static switching circuits, the terminals of individual units and outgoing user's connections should be readily accessible for checking and testing.

3. Because of the absence of moving parts, it is not possible (as in the case of contactors and relays) to determine by visual inspection whether or not a logic element is producing an output when the proper input circuits are energized. Signal lights or similar visual indicators are recommended to assist service people in checking the operation of a static switching system.

BIBLIOGRAPHY

1. "Cypak, A New Concept in Industrial Control," *Westinghouse Engineer*, July 1955.
2. W. G. Evans, W. G. Hall, and R. I. VanNice, "Magnetic Logic Circuits for Industrial Control Systems," *AIEE Paper* 56-91, *Applications and Industry*, July 1956.
3. J. P. Baker, "Cypak Systems, An Application of Logic Functions," *Westinghouse Engineer*, July 1956.
4. J. C. Ponstingl, "Static Electrical Controls," *Machine Design*, November 29, 1956.
5. R. A. Mathias, "Static Switching Devices," *Control Engineering*, May 1957.
6. R. A. Sitts, "Another Approach to Static Control," *Electrical Manufacturing*, June 1957.
7. J. Sheets and R. A. Brown, "Switching—Logic Functions," *Product Engineering*, August 1957.
8. D. L. Pierce, "Application of Static Switching in the Steel Industry," *Iron and Steel Engineer*, November 1957.
9. E. L. Rudisill, "Static Control," *Machine Design*, October 2, 1958.
10. J. W. Stuart, "Static Switching Control Design Considerations," *Electrical Manufacturing*, October 1958.
11. J. W. Stuart and R. A. Manning, "Logic Design Techniques of Static Switching Control for Transfer Machines," *AIEE Paper* CP58-1284.
12. F. L. Bosch, A. J. Fanthrop, and J. W. Stuart, "Static Control in Automatic Warehousing," *AIEE Paper* 58-1285.

13. G. J. Hess, "Static Control for Wood Chip Chemical Digester for the Paper Industry," *Electrical Engineering*, November 1958.
14. J. J. Duffy and S. Spatz, "Static Switching Today," *Electrical Manufacturing*, May 1959.
15. E. L. Rudisill, "Designing Static Control Circuits," *Machine Design*, October 1, 1959.
16. C. F. Meyer, "Packaged NOR-Logic Control," *Product Engineering*, January 4, 1960.

6 MOTOR PROTECTION

Although the principal purpose of a controller is to obtain specific performance, such as starting, acceleration, control of speed, retardation, and the like, certain protective features are also included as part of controllers to prevent damage to the motor resulting from unusual operating conditions. Foremost of those protective features is the protection of the motor against overloads which would cause damage to the motor from overheating, and which may create a fire hazard because of a motor burnout. The National Electrical Code makes it mandatory to provide motor overload protection, and practically all motor controllers include overload protection. Other protective features may be included as optional items.

EFFECT OF MOTOR OVERLOAD

An overload on a motor can be defined as an operating condition, not necessarily a short circuit, which causes a current in excess of normal current to flow in the motor power circuit. Thus overload protection can be defined as the protection of the motor, the control apparatus, and the branch-circuit conductors against excessive heating due to motor overload. It can be further defined as the effect of a device to cause interruption of current flow to the motor operating under an overload.

The damaging effect which an overload may have on the motor and the connected apparatus is caused by the excessive temperature which the increased current flow causes the apparatus to attain. Temperature rise of a motor increases approximately as the square of the load current. Thus each 10 per cent rise in load is accompanied by an increase of approximately 20 per cent in temperature. The attainment of excessive motor temperature is a function of the magnitude and duration of the overload and the ambient temperature in which the motor operates. In an extreme case, the temperature may rise so high as to cause melting of conductors, causing permanent damage to the motor, the control apparatus, or the motor branch-circuit conductors, and thereby constituting a fire hazard. But even if the tem-

EFFECT OF MOTOR OVERLOAD

perature does not rise to such an extreme value, any increase in temperature above that on which motor rating is based reduces the life of the insulation and results in shortened service life of the motor. For Class A motor insulation the motor life is halved by each 8 to 10 C increase in motor operating temperature.

The most commonly encountered causes of overload on a motor are:

1. Sustained overload caused by abnormal mechanical load on the motor shaft, or by low line voltage.

2. Too rapid duty cycles on intermittent drives, such as too frequent starting and stopping, which cause the rms value of current to exceed the rated motor current.

3. Excessive mechanical load, which causes the motor to stall or fail to start, thus drawing starting current for a considerable period of time.

4. Single-phasing of polyphase motors, preventing the motor from starting or causing excessive current to flow while the motor runs.

Although not an overload in the strict sense of the word, excessive ambient temperature in which the motor operates has an effect similar to a sustained overload. When the ambient temperature is higher than normal, the total temperature which the motor attains while carrying normal load is higher than the temperature on which the motor rating is based. This higher total temperature causes deterioration of the motor insulation in the same manner as if the motor were carrying a sustained overload while operating in normal ambient. Therefore, an ideal overload protective device would recognize not only the load on the motor but also the ambient temperature in which the motor operates.

To afford full protection to a motor, the tripping characteristic of the overload protective device should match the motor heating characteristic. To obtain the highest degree of service from the motor, the protective device should permit the motor to carry an overload for as long a time as is necessary for the motor to reach its maximum permissible temperature, but it should disconnect the motor as soon as that temperature is reached. Any overload protective device should thus have an inverse-time tripping curve, which should follow the motor heating curve as closely as possible.

No standards exist for the heating characteristic of electric motors, and considerable differences may exist between motors of the same horsepower rating but of different manufacture. An idea of the time in seconds which are required for general-purpose induction motors to

MOTOR PROTECTION

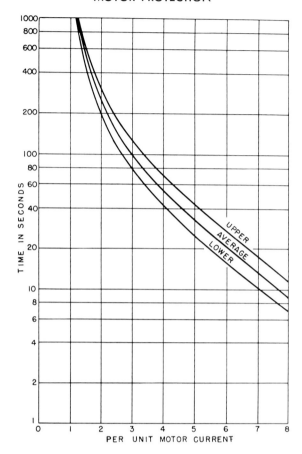

Fig. 1-6 Heating curve of typical general-purpose induction motors, giving time during which motor attains maximum permissible temperature for various amounts of motor load currents.

attain a maximum permissible temperature for various amounts of load current may be gained from Fig. 1-6. Three curves are plotted: an upper and a lower limit, and an average curve. While not every motor may be expected to fall within the band indicated in Fig. 1-6, most general-purpose induction motors will fall within this band. It has been plotted under the assumption that the motor will be permitted to reach a maximum temperature of 125 C while running and 150 C when stalled.

Motors must be protected not only against sustained overloads of

INHERENT PROTECTION 247

moderate magnitude, but they must also be protected against stalling. A permissible stall time of 20 seconds, with an average stall current of six times normal motor full-load current, is a generally accepted landmark which is followed by motor and control designers for correlating motor heating and adequate design of overload protection. It must be recognized, however, that the permissible stall time may change with changes in basic motor design. Advances in insulating materials which would permit higher motor temperatures and higher electrical loading of the active copper and iron in the motor may result in a reduction in the permissible stall time. Future motor design practices may change this 20-second landmark, which would require corresponding changes in the design of protective devices. While the principles of motor protection discussed in this chapter have universal significance, the numerical values of present-day relay characteristics may have to change in the future in order to accommodate future motor designs.

Motors for use with high-inertia loads may have to be designed for longer permissible stall time in order to tolerate the required long starting time. Such motor applications should be checked, and special designs of overload relays may have to be used in order to avoid premature tripping of the protective relay, thereby preventing the motor from starting.

Another motor application requiring special treatment is the protection of hermetic motors for use on hermetically sealed compressor drives for air-conditioning and refrigerating units. The principal difference between hermetic and general-purpose motors is the smaller ratio of stall current to normal running current. Protection of hermetic motors is described later in this chapter.

INHERENT PROTECTION

Since it is the objective of motor protection to prevent overheating of the motor windings, the ideal motor protector would be a device sensitive to winding temperature. It would not be necessary for such a device to be actually at the same temperature as the motor winding as long as the protector would recognize at all times a temperature which is truly proportional to the temperature of the motor winding. Because of limitations in the transfer of heat from the motor winding to the temperature-sensitive device, it has not been found practicable to design temperature-responsive protectors which could be used universally for any type of motor. However, direct response to motor temperature is a principle which has been used very successfully in certain areas. Since temperature-responsive protectors have to be

MOTOR PROTECTION

closely correlated with the motor and furnished as a part integral with the motor, temperature-responsive devices are called "inherent protectors." The accuracy of the response of an inherent protector depends on proper control of heat transfer between the motor and the protector. For this reason it is not practicable for the motor user to purchase such a protector as a supply item and assemble it on the motor himself. The protector must be designed specifically for the motor and assembled as part of it at the manufacturer's factory.

One of the oldest types of inherent protectors is the "resistance temperature detector" (RTD), which makes use of the change in resistance of a metallic resistance element as a function of temperature. A metallic resistor, having a positive temperature coefficient, is imbedded in a stator slot in intimate contact with the motor stator winding. This resistor is connected in a bridge circuit, and a voltage-responsive relay detects an unbalancing of the bridge as a result of change in resistance of the RTD. By proper calibration, the relay can be made to respond and shut down the motor when the temperature of the motor winding exceeds a permissible value of, say 125 C. When a motor is subjected to comparatively slowly changing sustained overloads, the RTD follows the temperature of the motor winding very accurately, thus permitting the motor to carry its load but shutting it down when the overload causes overheating of the motor.

There are two limitations to the usefulness of the RTD which have to be recognized. Since the resistance element has to be assembled in a slot of the machine, it takes up valuable winding space and is not a practicable protector for small and medium-size motors. Therefore RTD's are used only on large machines. Another limitation is the time lag between a change in motor-winding temperature and a corresponding change in the resistance of the RTD. Therefore they do not provide stall protection for motors because, with the high stall current, the motor winding would be damaged before the RTD responds. Stall protection has to be provided by some other means.

Results similar to an RTD are obtained by the use of a thermostat responding to motor temperature. One form of thermostat, which has found practical application in this country and abroad, is a thermostat utilizing a Spencer disk. This disk consists of bimetal, and it is shaped so that it is slightly convex at normal temperature. When heated to a certain temperature, the Spencer disk snaps into a slightly concave shape. When the disk is cooled, it snaps back into the original shape. This snapping movement, which takes place at a definite temperature within narrow limits, the magnitude of which is determined by the size

INHERENT PROTECTION 249

and prestressing of the disk, can be utilized for the actuation of contacts. Such a thermostat does not need to be imbedded in a slot of the machine, but may be attached to the end turns of the stator winding or to the stator frame. Thermostats intimately bonded to the motor winding or the frame respond quite accurately to overheating of the motor as a result of sustained overloads. Compared with RTD's, thermostats have the advantage that they do not take up winding space. However, they also have the limitation that they do not respond fast enough to afford protection of the motor against a stall.

This disadvantage has been overcome by the Klixon * protector. A Klixon protector consists of a Spencer disk and a heater, through which motor load current flows. As long as a current equal to or less than normal load current flows through the heater, the amount of heat thrown into the Spencer disk has very little effect on the temperature of the disk. However, when a motor stalls and its current rises to, say six times normal motor current, the heat contributed by the heater is increased thirty-sixfold, which results in rapid heating of the disk. By proper correlation of the design of heater and disk, response of the disk may be speeded up to the point where the disk responds before the motor has attained a dangerously high temperature during a stall.

Klixon protectors have been used for many years in large quantities for the protection of fractional-horsepower appliance motors. On these protectors the Spencer disk actuates a contact which is connected directly in the motor power circuit, so that when the disk snaps, line current of the motor is interrupted. Since the protector is current sensitive, each motor and its protector have to be carefully correlated. For this reason, the Underwriters' Laboratories, Incorporated, test motor and protector as a unit, and they are listed as a unit. Protectors are not tested and listed as supply items for general use.

Klixon protectors have also been developed for use with three-phase industrial motors. They can be obtained for use with squirrel-cage motors rated up to $7\frac{1}{2}$ horsepower at 220 or 440 volts. Figure 2-6 illustrates a Klixon protector for use with integral-horsepower three-phase motors. A is a view of the assembled protector. It is encased in a molded plastic housing with solder-type terminals brought out for connection to the motor winding. B is a cross section of a protector. The Spencer disk carries the movable contacts, and the heater is visible below the disk. C is a view of a general-purpose motor, illustrating

* Registered trade mark of the Metals and Controls Corporation.

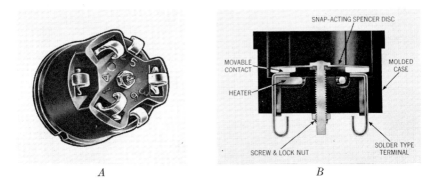

Fig. 2-6 Klixon protector for integral-horsepower three-phase squirrel-cage motor. *A*. Assembled protector. *B*. Cross section of protector. *C*. General-purpose motor with protector attached to stator laminations. (*A* and *B*, courtesy Metals and Controls Corporation.)

INHERENT PROTECTION

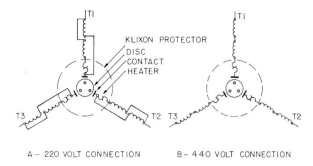

Fig. 3-6 Connection of Klixon protector in wye point of dual-voltage squirrel-cage motor.

the manner in which the protector, enclosed in a cap, is attached to the stator frame near the end turns of the winding.

Connections of a Klixon protector to a wye-connected dual-voltage motor are indicated in Fig. 3-6. The protector is connected in the wye point of the motor. The disk forms the wye point, and power flow through the motor is interrupted when the disk snaps and opens the wye point. The heater is connected in the motor circuit in such a manner that the same current flows through the heater with the motor winding connected for either 220- or 440-volt operation. By proper coordination of motor and protector, it is possible to limit motor temperature to 125 C for running overloads and to 150 C for motor stalls. These temperature limits for motor and protector combinations have been established by Underwriters' Laboratories, Incorporated, tests. Klixon protectors may be obtained with automatic resetting, that is, with contacts that close automatically when the protector cools after the motor shuts down. Protectors can also be obtained with manual reset, which requires the pressing of a button to reclose the contacts after the protector has tripped open. A protector with automatic reset will automatically cause the motor to restart after a shutdown, and should only be used where automatic restarting does not endanger the operator.

Klixon protectors are particularly suited for the protection of motors in hermetic compressor units, which are used in refrigeration and air-conditioning equipments. In these units the motor is inside the cooling system, and it is cooled by the refrigerant. Therefore the temperature of the motor is not a function of its loading, because high load on the motor may be associated with low temperature of the refrigerant. A temperature-responsive protector actually "sees" the motor temperature, and thus safeguards the motor if anything should go wrong with the compressor or refrigerating system.

Since the protector does not disconnect the motor from the line, a disconnecting means, such as a switch, a circuit breaker, or a contactor, must be provided to isolate the motor from the line when the equipment is shut down. It should also be considered that Klixon protectors can only be used on three-phase motors with wye-connected stator winding. They cannot be used on motors with delta-connected stator winding. Another point to consider is the accessibility of the motor. If automatic restarting is objectionable from the point of view of safety, then the motor must be in an accessible location to permit resetting the protector by the manual reset button. While protection of motors against single-phasing will be treated in detail later in this chapter, it is to be noted that, under single-phasing conditions, Klixon protectors shut down the motor at a hot spot temperature which is somewhere between 10 C and 15 C above the temperature at which shutdown occurs for three-phase operation.

CHARACTERISTICS OF OVERLOAD RELAYS

A small number of industrial motors are protected by magnetic overcurrent relays. These are mostly motors used on intermittent drives for which motors are rarely selected on the basis of their thermal capacity, but generally on the basis of torque requirements. The majority of overload relays used on industrial motors are thermal overload relays. They use the heat storing ability of a thermal-responsive element and connected mass. The heat generated by the motor load current flowing through the relay is conducted, convected, or radiated to this element which causes the relay contact to be actuated when it has reached a certain temperature. By proper design of the relay parts and by providing the necessary heat-storing capacity, the heating characteristic of the relay can be made to follow closely the heating characteristic of the motor. Because the heating characteristic of such relays simulates the heating characteristic of the motor, the heat-generating and storing portion of such relays constitutes a replica of a motor. Hence thermal overload relays are called "replica-type relays." Ideally, a replica-type relay actuates its contact just prior to the point at which the motor reaches its maximum permissible temperature. As a consequence, a replica-type relay permits the motor to carry limited overloads for a limited

CHARACTERISTICS OF OVERLOAD RELAYS

period of time, but acts to disconnect the motor from the line when the motor is in danger of overheating.

Most overload relays used on industrial controllers have normally closed contacts; that is, the contacts are closed when no current is flowing and open when the relay responds to an overload current. Contactors used on magnetic controllers are generally closed by holding their coil energized. Overload protection is obtained by the overload relay contacts interrupting the line-contactor coil circuit, causing the line contactor to drop out and disconnect the motor from the line. Contrary to control practice, switchgear practice prefers mechanically latched-in circuit breakers. A trip coil, energized on overload, causes the breaker to open. Thus overload relays used on switchgear-type motor starters generally have normally open contacts which close and energize a trip coil when an overload occurs.

Contacts of overload relays may be reset automatically or by hand. On automatically reset relays, the contacts reclose automatically after the motor has been disconnected from the line. On hand-reset relays the contacts stay open until they are reclosed mechanically, usually by pressing a reset button. On magnetic controllers using momentary-contact control-circuit devices or master switches with undervoltage protection feature, either type of relay can be used. Although the hand-reset feature is not necessary to obtain satisfactory operation, many users prefer hand-reset overload relays because the operator is forced to go to the control panel to reset the relays. This manual operation prevents the operator from restarting his drive immediately upon an overload, and there is a greater probability that the operator will investigate the cause of the overload. Hand-reset overload relays must be used on controllers which are started and stopped with maintained-contact control-circuit devices. Automatically reset relays would cause the motor to "pump"; that is, the motor would continually attempt to restart against the overload.

The question then arises how many relay units should be used and how they should be connected in the motor circuit. Overload relays may be built as single-unit relays; that is, the relay is connected in one line to the motor and actuates its own contact. Overload relays may also be built as multiunit relays in which several actuating units, connected in several lines to the motor, actuate a common contact. This discussion of number of overload relays required is based on the assumption that single-unit relays are used. It should be understood that one single relay containing multiple units obtains equivalent per-

MOTOR PROTECTION

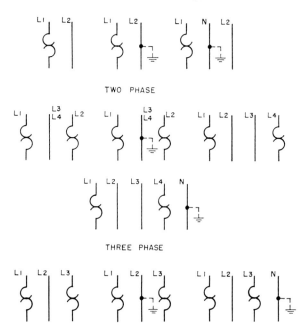

Fig. 4-6 Standard connections of overload-relay units for use with single- and polyphase motors.

formance. As an example, two single-unit relays are equivalent to one two-unit relay.

Standard connections of overload relays in single-phase and polyphase motor lines, as specified in the National Electrical Code, are indicated in Fig. 4-6. One relay in an ungrounded line is used for the protection of single-phase and direct-current motors. Two relays are used for the protection of two-phase motors, one relay being connected in one ungrounded line of each phase for motors connected either to a three- or four-wire system. Two overload relays connected in two ungrounded lines are used to protect three-phase motors. Two relays provide adequate overload protection for two-phase and three-phase motors if the motors are single-phased by the loss of one line to the motor stator terminals. While the number of overload relays indicated in Fig. 4-6 obtains adequate protection in the great majority of applications, certain conditions of single-phasing may occur under which two overload relays may not provide complete protection to the

BIMETAL OVERLOAD RELAYS

motors. These conditions are discussed in detail under the heading "Single-Phasing."

BIMETAL OVERLOAD RELAYS

Bimetal overload relays use a heat-responsive element consisting of a strip of bimetal which is deflected by heat. A typical bimetal relay is shown in Fig. 5-6. The working parts of the relay are assembled in a molded case, one cover of which is removed to show the assembly of parts. When the bimetal strip deflects, it releases a latch, and a snap-action mechanism causes the relay contact to open. Thus the mechanical work done by the bimetal strip is kept low, so that a minimum of mechanical bending force is exerted on the strip, preventing the strip from becoming deformed upon repeated heating and cooling. When the bimetal strip has cooled after a tripping, the relay may be reset manually by pressing the reset arm, which closes the contact and latches the snap-action mechanism through the insulated operator. An adjustment knob is provided which permits the shifting of the position of the bimetal strip. Turning the knob changes the distance

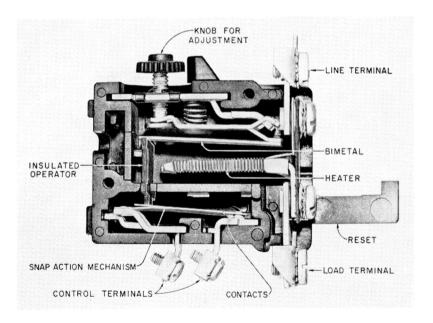

Fig. 5-6 Resistance-type thermal overload relay with bimetal element. Side view with cover removed.

MOTOR PROTECTION

through which the bimetal strip has to travel before unlatching the contact mechanism. By this adjustment, the trip current to which the relay responds can be changed by approximately ±15 per cent. Motor current flows through a resistance-type heater. The heat generated by the current flow is radiated, and to a limited extent conducted and convected, to the bimetal strip which forms a cantilever paralleling the heater. The heater can be readily removed by loosening two screws. By interchanging heaters of various sizes, it is possible to use the relay for motors of different full-load currents.

For those applications for which an automatically reset relay is preferred, a simple adjustment of a reset spring on the outside of the relay permits to change contact operation from manually reset to automatically reset. After the relay has tripped, a period of time elapses before the contacts reclose, which is the time required for the bimetal strip to cool and return to its original position.

Figure 6-6 shows a typical tripping curve of a resistance-type bimetal overload relay working in a normal 40 C ambient temperature. In judging the performance of thermal overload relays, it must be considered that these relays are not precision-type relays as one would expect to find on switchgear in connection with power system relaying. Thermal overload relays are mass-produced devices on which the designer strives to obtain optimum performance at minimum cost, permitting the use of such relays on comparatively inexpensive motor starters.

In Fig. 6-6, the tripping characteristic is given as a band. It indicates that the tripping curve of any relay of a given type, when used with any size heater available for that relay, will fall within that band. One reason for this tolerance band is the unavoidable mechanical tolerances in the manufacture and assembly of structural parts which have a noticeable effect on the distance the bimetal strip has to travel in order to release the contact mechanism. A second reason for the tolerance band of the tripping curve is variation in the thickness of the bimetal strip, resulting in varying amounts of heat required to deflect the bimetal the desired amount. A third cause of variation are the tolerances to be expected in the heater itself. Variations in the composition and cross-section of the resistive material cause variations in resistance, and consequently in the amount of heat generated by a given amount of current. When applying thermal overload relays, it is necessary to take this tolerance band into consideration. It is not sufficient to select an overload relay on the basis of an average curve;

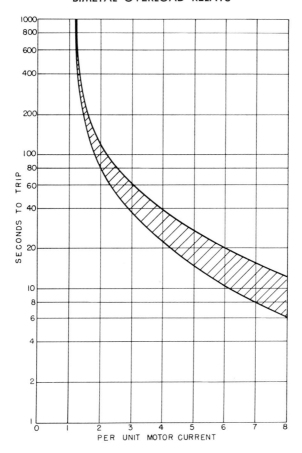

Fig. 6-6 Tripping curve of a resistance-type thermal overload relay, used with a general-purpose 40 C rise motor.

the possible deviations from that average must be evaluated when close correlation between motor and relay is required.

Strictly speaking, the tripping characteristic of Fig. 6-6 applies to a specific type of overload relay which has been designed for use with general-purpose squirrel-cage motors having a service factor of 1.15 and a permissible stall time of the order of 20 seconds. The relay will not trip on 1.15 times motor full-load current, and at a stall current of 6 per unit the tripping time is between 10 and 20 seconds. While other types of thermal overload relays offered for the protection of general-

MOTOR PROTECTION

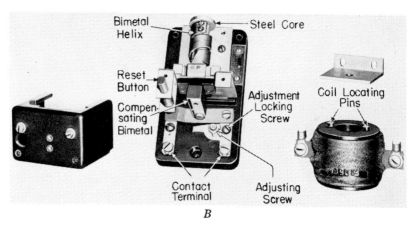

Fig. 7-6 Induction thermal overload relay, temperature-compensated. *A*. Assembled relay. *B*. Exploded view showing component parts.

BIMETAL OVERLOAD RELAYS

purpose motors may deviate slightly from the characteristic shown, they follow the same tolerance band quite closely, and this curve can be accepted as a reasonably accurate general curve for resistance-type overload relays.

Relay heaters have a limited heat-storage capacity. There is a time lag between the heating of the resistance heater and the heating of the bimetal strip. Heaters may melt and damage the relay permanently if subjected to fault currents for any length of time. Thermal overload relays do not provide short-circuit protection. Fuses or circuit breakers must be included in the motor circuit to prevent damage to the overload relays under fault-current conditions. Proper coordination between overload relays and short-circuit protection means are discussed in detail in Chapter 7.

Another method of getting heat into the bimetal strip is by induction, and a relay based on that principle is shown in Fig. 7-6. A is a view of the assembled relay, and B is an exploded view showing the component parts of the relay. The relay has a series coil which is connected in the motor circuit. Various coils are available for use with motors of different full-load currents. Mounted within the coil is a steel core over which is assembled the bimetal strip, wound in the shape of a helix which is short-circuited within itself. This bimetal helix acts like a short-circuited secondary of a transformer. The current induced in the helix causes it to become heated and to rotate, thereby opening the relay contact. After a trip, the relay contact has to be reset by hand, after the helix has cooled down and returned to its original position. By means of an adjusting screw the helix travel required for tripping the contact can be varied. Thus the relay trip current can be adjusted between 90 and 110 per cent of its nominal trip current.

Induction overload relays are inherently more expensive than the previously described resistance-type relays. For this reason induction relays are not frequently used for the protection of the smaller general-purpose motors, but are mostly used on larger motors and in the secondaries of current transformers on controllers for high-voltage motors. These larger motors are often designed especially for the application on which they are to be used. Depending on starting requirements, the permissible stall time may vary considerably. By using different assemblies of iron core and bimetal helix, relay forms are obtained which have different tripping characteristics. Typical tripping curves of three forms of induction overload relays are shown in Fig. 8-6. A pertains to a fast-trip relay. The stalled-rotor current of six times normal motor current obtains an average relay trip time of

MOTOR PROTECTION

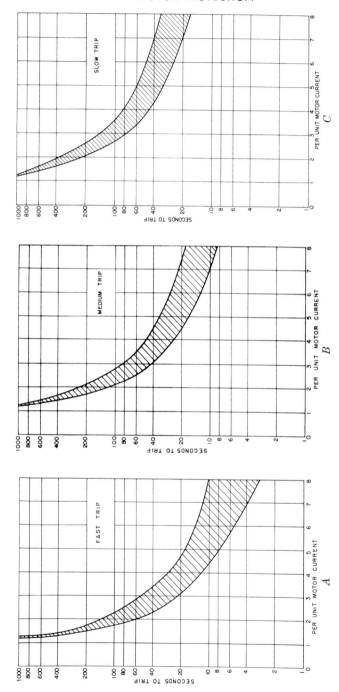

Fig. 8-6 Tripping curves of induction thermal overload relays for different tripping ranges. *A.* Fast trip. *B.* Medium trip. *C.* Slow trip.

SOLDER-FILM OVERLOAD RELAYS

the order of 10 seconds. B is the characteristic of a medium-trip relay on which the average trip time at stalled-rotor current is approximately 20 seconds. This relay has a characteristic comparable to that of a general-purpose resistance-type relay. C applies to a slow-trip relay for use with motors especially designed for driving high-inertia loads requiring a long starting time. The average tripping time at stalled-rotor current of this relay form is about 30 seconds.

Since the tripping point of the overload relay is determined by the total temperature of the bimetal element, the tripping characteristic of the relay is affected by the ambient temperature in which the relay is installed. Overload relays may be installed in enclosures within which a certain amount of heat is generated. An example are group controllers, such as motor control centers serving a multiplicity of motors, the starters of which are in a common enclosure. Load changes may cause considerable change in the temperature inside the enclosure, namely the ambient temperature within which the relays operate. Overload relays may be installed in locations subjected to pronounced seasonal changes in the temperature of the air surrounding the controller. These environmental variations would result in noticeable changes in the current or time within which the relays trip. To compensate for these ambient temperature changes and to make the relay tripping characteristic depend on current only within a certain range of temperature variations, the relay of Fig. 7-6 is "ambient-compensated." A compensating bimetal strip is included which is so arranged that temperature changes produce a deflection nearly equal to that of the bimetal helix, and in the opposite direction.

SOLDER-FILM OVERLOAD RELAYS

Another type of thermal overload relay which has found widespread practical application utilizes a film of eutectic solder as the heat-responsive element. Eutectic solder is an alloy with a low melting point. It changes abruptly from a solid to a liquid state and vice-versa at some critical temperature. Figure 9-6 illustrates the principle of solder-film relay operation. A spindle with a ratchet wheel is held in a sleeve by the eutectic solder. A pawl, which is loaded by an operating spring, is held in position by engagement with the ratchet. A heater wound in the shape of a coil is slipped over the sleeve. When sufficient heat has been generated by the relay heater, the solder film softens suddenly, releasing the spindle and the pawl, which, under the influence of the operating spring, moves to trip the contact unit. After the relay has cooled down sufficiently to permit the solder to harden

MOTOR PROTECTION

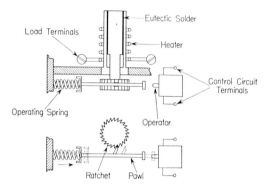

Fig. 9-6 Illustrative diagram of solder-film-type thermal overload relay.

and lock the spindle in position, the relay may be reset manually by pushing back the pawl and engaging it again with the ratchet. This type of relay can only be reset manually. The tripping characteristic of such a relay is comparable to that of a bimetal relay, as indicated in Fig. 6-6.

Typical solder-film overload relays are shown in Fig. 10-6. A is a completely assembled relay with heater in place. The side plate is turned up to show the assembly of internal parts of the relay. The contact mechanism with its actuating spring, the ratchet, and the pawl can be readily recognized. In this view, the relay on the left is shown in the tripped position, and the relay to the right in the reset position. B is a view of a relay with the heater removed. The sleeve and spindle assembly can be removed; and by interchanging spindles of various dimensions and mass, the relay trip characteristic can be modified. It is possible to obtain a fast-trip, medium-trip, or slow-trip characteristic similar to the ones shown in Fig. 8-6. C shows a design in which the heater and the solder-film heat-responsive element are assembled into an integral unit. Heater and solder pot are assembled on the relay contact unit as one complete assembly. This design permits close control of tolerances in relative position of heater and solder film during factory assembly, thereby increasing the accuracy of the relay tripping characteristic.

APPLICATION OF OVERLOAD RELAYS

A certain type of thermal relay can be used for a considerable range of motor currents. The specific current rating, which determines the motor with which a relay can be used, depends on the size of the heater

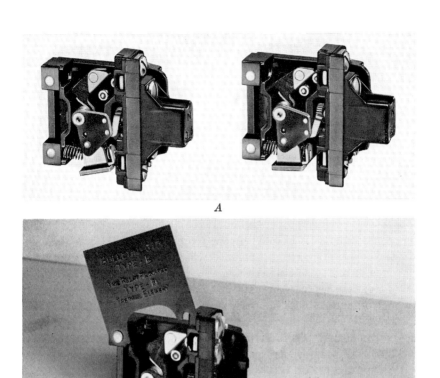

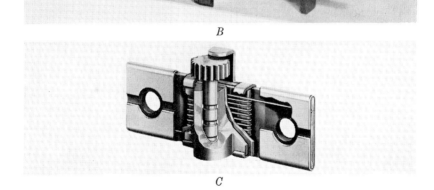

Fig. 10-6 Solder-film thermal overload relays. *A*. Assembled relay with heater. *B*. Relay without heater, showing interchangeable spindle. *C*. One-piece construction of heater and heat-responsive element. (*A* and *B*, courtesy Allen-Bradley Company; *C*, courtesy Square D Company.)

employed. For each relay, the manufacturer lays out a line of heaters (or coils for induction relays), which cover the current range over which the relay is to be used. In order to satisfy the requirements of the National Electrical Code and the Underwriters' Laboratories, lines of heaters or coils are laid out such that the spread between individual heaters is not greater than 10 per cent. For each relay heater the design engineer establishes a nominal current rating at which the relay will eventually trip when mounted in the open by itself in an ambient temperature of 40 C. This nominal current rating is the basis on which a design engineer establishes the motor current rating for which the relay can be used on a specific controller. Since thermal relay performance is affected by the temperature of the air surrounding the relay, it makes a difference whether the relay is mounted in the open or in an enclosure. Furthermore, the heat transmitted to the relay from the current-carrying parts of a starter has an influence on the performance. For this reason, it is customary for control manufacturers to specify the motor current with which the overload relay heater can be used when forming part of a specific starter. Relay tables are then consulted by the user to select the proper relay heater for use with a specific motor.

It would be economically impossible to design a heater or coil for every conceivable value of motor current with which a relay could be used. Therefore, manufacturers' relay tables list a range of motor currents for each heater or coil. General-purpose motors with Class A insulation, rated for continuous duty at 40 C rise, have a service factor of 115 per cent. The National Electrical Code specifies that overload relays be so selected that the ultimate trip current of the relay is not higher than 125 per cent of motor current. Hence the difference between 115 per cent service factor current and 125 per cent of maximum permissible relay trip current constitutes the maximum 10 per cent steps between relay heaters which manufacturers' tables adhere to.

When thermal overload relays are applied to motors which do not have a 115 per cent service factor, the National Electrical Code requires that the nominal relay current must not exceed 115 per cent of rated motor current. Heaters selected on that basis work out satisfactorily when used for continuous-duty motors with Class A insulation rated on a 50 to 55 C rise basis. However, motors rated on a short-time duty basis would not be properly protected by such relays. Table 8 is an aid in selecting overload relay heaters or coils for use with motors rated on a basis different from the standard general-purpose continuous-duty 40 C motor rating. Rated nameplate current of

APPLICATION OF OVERLOAD RELAYS 265

Table 8. Conversion Factor for Selecting Overload Relay Heaters from Table for 40 C Rise Motors with Class A Insulation

Motor Rating Basis	Continuous Capacity of Motor in Terms of Rating	Select Heater for Current Given in Relay Table, Multiplied by
40 C rise, continuous	115%	1.0
50–55 C rise, continuous	100%	0.9
50–55 C rise, 1 hour	80%	0.8
50–55 C rise, 30 minutes	70%	0.75
50–55 C rise, 15 minutes	55–60%	0.70

the motor is multiplied by the factor indicated in the last column of this table, and that current is used for selecting the relay heater from a standard general-purpose relay table. Motors with Class B or F insulation may have heating characteristics differing from motors with Class A insulation. Manufacturers' recommendations should be followed for proper relay heater selection for such motors.

Current rating of motors as well as thermal overload relays is based on operation in an ambient temperature of 40 C. Since the total temperature determines the deterioration of motor insulation as well as the tripping of the overload relay, special consideration must be given to the effect of ambient temperature on motor and relay performance. In Fig. 11-6, the percentage of rated current which produces a limiting total temperature is given as a function of ambient temperature for running motors, for nonambient-compensated thermal relays on open starters and for like thermal relays on enclosed starters. Because of the effect of windage, a running motor is capable of carrying a higher percentage of current than the overload relay. Owing to the effect of free air circulation, a relay on an open starter will trip on a higher current than a relay on an enclosed starter in the same ambient. If it is known that overload relays are installed in locations with higher than normal ambient, relay heaters should be selected for a higher current than called for in the standard relay heater table. Otherwise the overload relays would trip on currents less than the full-load current of the motor, thus preventing the motor from carrying loads commensurate with the motor's capabilities. Higher than normal ambient temperatures may be encountered in certain locations in industrial plants

MOTOR PROTECTION

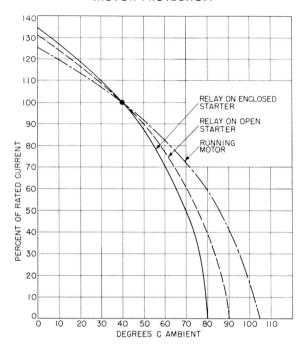

Fig. 11-6 Effect of ambient temperature on thermal relay and motor performance, indicating percentage of rated current obtaining the same temperature with varying ambient temperature.

close to furnaces and other heat-producing equipment. Another reason might be exposure of the control equipment to extremely high summer temperatures, and in some cases the temperature inside of an enclosed controller may be considerably higher than room temperature. On group controllers containing a multiplicity of motor starters in a confined space, as for instance in a motor control center, the inside temperature in which the overload relays have to operate may be found 15 to 20 C higher than the surrounding room temperature.

Whereas high ambient temperature may result in nuisance tripping of overload relays, but not in motor burnout (unless relay heaters are selected for too high a current), operation in below-normal ambient endangers the motor. Figure 11-6 indicates that below 40 C the motor attains its limiting temperature faster than the relay would trip, and burnout under overload or stall may result.

While Fig. 11-6 applies to a resistance-type bimetal relay without ambient compensation, the shape of the curves are typical for any

APPLICATION OF OVERLOAD RELAYS

uncompensated thermal relay. Where such relays are to be installed in locations in which ambient temperature deviates greatly from 40 C, the relay manufacturers should be consulted if close protection of the motor is a critical requirement.

When relays are applied in locations subjected to widely varying ambient temperatures, the use of ambient-compensated relays would be indicated if such use is economically feasible. On ambient-compensated relays the variation in current which eventually causes relay tripping varies approximately 3 per cent for a variation of 10 C in ambient temperature. This relationship can be assumed to hold over a temperature range of approximately 20 to 80 C. While these figures are indicative of the order of magnitude of compensated overload relay variation, the relay manufacturer should be consulted on specific data for any accurate relay application study.

The variation in tripping time of an ambient-compensated relay may even be too great to avoid nuisance tripping at high ambient temperature, and still obtain reliable motor protection at the lower end of the ambient range. This condition may exist if a critical process drive motor is controlled by a controller installed outdoors, where the relay is subjected to wide seasonal variations in ambient. A possible solution is an induction-type overcurrent relay with an inverse-time tripping characteristic properly selected to match the motor capability.

By far the majority of industrial motors are used on drives with comparatively low inertia, and the starting time of such drives is only a few seconds. In most cases the heat generated in the overload relay by the starting current is considerably less than that required to trip the relay, so that starting the motor is generally no problem as far as selection of a suitable overload relay heater is concerned. Permissible stall time of general-purpose squirrel-cage motors is of the order of 20 seconds, which means that a motor is capable of tolerating a 20-second starting time infrequently. Because of the variations in overload relay trip characteristics, it cannot be assumed that any relay selected from a standard table will permit a 20-second accelerating time.

Another factor which needs to be considered is the overshoot of indirectly heated thermal overload relays. In any indirectly heated relay, there is a temperature difference between the heater and the thermal-responsive element. When current flow to the heater is interrupted, heat continues to flow to the thermal-responsive element; and when its temperature is close to the trip temperature, the relay may trip after current flow through the heater has ceased. The degree of overshoot varies with different relay types and also with different heater

MOTOR PROTECTION

sizes. It is entirely possible that a relay will trip when motor starting current flows for approximately 70 per cent of the permissible stall time after which the relay is expected to trip. On directly heated overload relays, such as induction-type relays, the amount of overshoot is much smaller. Such relays can be expected to be free from nuisance tripping when the starting time is as high as 95 per cent of the expected tripping time on stall current.

When standard thermal overload relays (having a nominal tripping time of 20 seconds at six times motor current) are applied on the basis of standard relay-selection tables, it is generally understood that such relays will not cause nuisance tripping during starting as long as the starting period does not exceed 10 seconds. When motors are used on high-inertia drives requiring a starting period of more than 10 seconds, motor capability and relay characteristic should be carefully checked to make certain that the load can be started without nuisance tripping. Relays with a slow-trip characteristic may have to be used. Motors designed especially for starting high-inertia loads have longer permissible stall times than the customary 20 seconds for general-purpose motors. After the drive has been started, the thermal characteristics of such motors are comparable to those of standard motors; and standard thermal overload relays are adequate for protecting such motors against running overloads. To avoid nuisance tripping during starting, overload relays may be connected in the secondary circuits of current transformers which are so selected that they saturate on starting current. As an example, a current transformer having a secondary current of four times motor full-load current with six times motor full-load current flowing in the primary, might be used with a standard thermal overload relay and still avoid nuisance tripping with a starting period as high as 30 seconds. When making a relay study for a high-inertia load with long starting time, it is particularly important to investigate carefully the effect which high ambient temperature may have on the relay trip characteristic.

The pointers on overload relay selection for motors having a longer than 10-second starting time, are based on the 20-second stall time which, at this writing, is the prevalent stall time to which general-purpose squirrel-cage motors are designed. If motor-design practices change and permissible stall times become shorter, a corresponding change will have to be made in the tripping characteristic of general-purpose thermal overload relays. However, the general considerations described in the preceding paragraphs would still apply.

OVERLOAD RELAYS FOR HERMETIC MOTORS

Special consideration needs to be given to the protection of motors used on hermetically sealed compressor units such as those generally used for refrigeration and air conditioning. A hermetically sealed compressor consists of a mechanical compressor driven by an electric motor, both of which are enclosed in the same housing. The motor operates in the refrigerant atmosphere. Squirrel-cage a-c motors are used on commercial equipments. Because they operate at a lower than the normal 40 C ambient temperature, they can deliver higher output and can draw higher current than a standard motor of equivalent frame size operating in the open, without exceeding the permissible total temperature of the motor windings. Hermetically sealed compressors have their nameplates marked with the current drawn by the motor while driving the compressor, and the stalled-rotor current.

Many hermetic motors, particularly in the smaller sizes, are equipped with inherent protectors which respond directly to the temperature of the motor windings. Inherent protectors offer the advantage that any cause of motor overheating, not only overloading but also any failure in the refrigerant circulation, will be detected. Inherent protectors are presently available only for motors of limited size, and no such units can be obtained for the larger motors. Therefore the larger compressor units have to be protected by overload relays.

Permissible full-load current of a hermetically sealed motor is higher than the rated full-load current of a standard motor of equivalent frame size. However, the stalled-rotor current of a sealed motor is approximately the same as the stalled-rotor current of a standard motor of comparable size. This means that the ratio of stalled current to normal running current of a hermetic motor is considerably less than the approximately six times which can be expected on a standard general-purpose motor having the same full-load current. In addition, hermetic motors have no service factor. If a standard thermal overload relay were selected from a standard general-purpose relay listing, it would have to be selected for a current rating corresponding to the compressor motor rated running current. Should such a motor stall, the overload relay would be too slow in tripping to prevent burnout of the motor. Therefore, special fast-trip relays have been developed with shorter tripping time on stall currents. To obtain the required short trip time, the heat-responsive element is heated directly by load current, thereby obtaining faster heating under stalled-current condition.

270   MOTOR PROTECTION

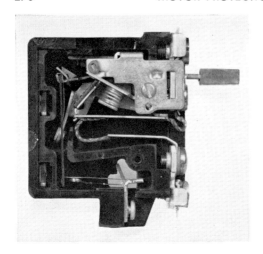

Fig. 12-6 Directly heated bimetal overload relay for protecting hermetic motors.

A bimetal relay for the protection of hermetic motors is shown in Fig. 12-6. Load current flows through a heater and also through the bimetal, thus generating a significant portion of the total heat directly in the bimetal strip. The corresponding trip characteristic is given in Fig. 13-6. A comparison with Fig. 6-6 indicates that the fast-trip relay at approximately three and a half times rated motor current obtains a tripping time comparable to that of a standard general-purpose relay at six times rated motor current.

OVERLOAD PROTECTION OF WOUND-ROTOR MOTORS

Wound-rotor motors do not differ significantly in stator design from squirrel-cage motors, and are likewise protected against running overload by standard general-purpose thermal overload relays. Motor temperature rating and service factor are considered in the same manner as for squirrel-cage motors. Since starting currents of wound-rotor motors are considerably smaller than for squirrel-cage motors, there is no problem in regard to nuisance tripping of the overload relay during the starting period, except possibly in rare instances where an extremely long accelerating period calls for a specially designed motor and correspondingly a specially selected overload relay.

When wound-rotor motors are likely to be stalled because of impact loads, with the slip rings short-circuited, precautions must be taken to remove the motor from the line immediately upon a stall. If a wound-rotor motor is stalled with no resistance in the rotor circuit, the stall current flowing through the slip rings at standstill would burn the

OVERLOAD PROTECTION OF WOUND-ROTOR MOTORS 271

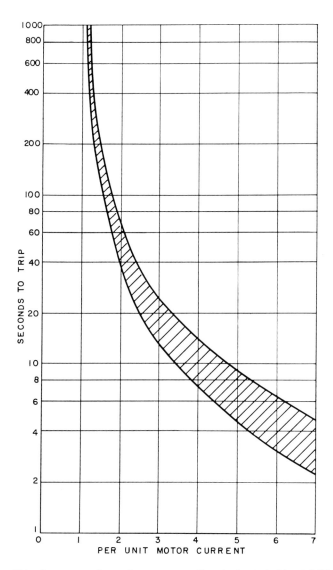

Fig. 13-6 Tripping curve of a resistance-type, directly heated, bimetal "fast-trip" thermal overload relay for protecting hermetic motors.

slip rings and damage the brushes. Instantaneous overcurrent relays should be used to trip out the line contactor. Since motor starting current and permissible overload currents are considerably below the stalled-rotor current, these instantaneous relays can be set sufficiently high to prevent tripping during starting and during normally expected load-current peaks. A current setting of the order of three times rated motor current should obtain satisfactory results in most cases.

OVERLOAD PROTECTION OF SYNCHRONOUS MOTORS

The stator winding of synchronous motors can also be protected by thermal overload relays of the same type as used for squirrel-cage motors. Protection against running overloads presents no special problems. When synchronous motors are used on high-inertia drives, as on flywheel motor generator sets, or for fans or compressors, the overload-relay tripping characteristic would have to be checked for possible nuisance tripping on starting.

Thermal overload relays in the motor stator circuit do not protect the squirrel-cage winding when the motor operates out of synchronism. Squirrel-cage protective relays, as described in Chapter 4, have to be relied upon for protecting the squirrel-cage winding on the rotor.

INSTANTANEOUS OVERLOAD RELAYS

For intermittent drives such as cranes and hoists, which are usually driven by short-time rated motors, generally of the wound-rotor type, the motor rating is usually established on the basis of torque requirement rather than thermal capability. In most instances, the motor does not require protection against overheating, but both motor and drive should be protected against excessive current and torque peaks which may be caused by impact loads or careless manipulation of the control by an operator. It is customary to include instantaneous overload relays on crane, hoist, and similar controllers.

Instantaneous overload relays are magnetically actuated overcurrent relays. The one shown in Fig. 14-6 has two coils, one for each of two phases of a three-phase motor, or the two phases of a two-phase motor. A plunger is drawn into the coil when the current exceeds the relay setting. The impact of the plunger opens the relay contact, which is generally connected to drop out the line contactor.

SINGLE-PHASING

Special problems arise when three-phase motors are single-phased because of opening of a line to the motor or a line in the power dis-

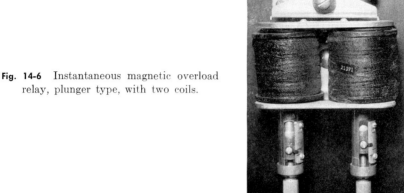

Fig. 14-6 Instantaneous magnetic overload relay, plunger type, with two coils.

tribution system. Consideration has to be given to the protection of three-phase motors under such conditions.

Single-phasing is a special case of unbalancing the voltage applied to the motor terminals. Any unbalanced three-phase system can be considered as a combination of a positive-sequence system and a negative-sequence system. Both of these systems cause currents to flow in the motor windings. Total heating of the stator is the sum of the heating due to positive-sequence and negative-sequence currents. Some additional heating is caused by changes in spatial distribution of the heating due to the presence of negative-sequence current. However, this effect can generally be neglected. An important cause of additional heating of the motor may be increased heating of the rotor resulting from negative-sequence currents. The ratio of negative-sequence rotor current to stator current is increased. Furthermore, motor resistance is increased because of skin effect in the rotor winding because of increased frequency of negative-sequence currents. As far as standard general-purpose motors in the smaller and medium frame sizes are concerned, this additional rotor heating generally does not lead to motor damage. Operating experience and motor damage records indicate that increased rotor heating due to negative-sequence currents is not an important factor in motor failures. However, in the case of larger motors with fabricated squirrel-cage windings with brazed or welded end rings, as well as in the case of wound-rotor motors, the additional rotor heating due to negative-sequence currents may lead to damage of the motor. On such motors, special precautions may have to be taken to disconnect the motor immediately upon single-phasing in order to prevent damage due to negative-sequence currents.

Three-phase distribution systems, as long as all three phases are intact, are generally free of troublesome unbalances. System operators maintain generally a high degree of balance between the three phase voltages. Public utility systems supplying combination light and power circuits have the single-phase loads carefully distributed on the feeders, and utilities are generally required through their power contracts to maintain a high degree of phase balance. In industrial plants, motors are generally connected to a 440-volt power system, whereas the single-phase loads, mostly lighting loads, are connected to separate lighting circuits operating at a different voltage. Also large single-phase loads which may exist in an industrial plant are usually connected to circuits separate from the motor circuits. Therefore, the degree of unbalance encountered on the usual three-phase power distribution systems is so small that its effect may be neglected as far as motor performance and motor protection is concerned.

A special case of voltage unbalance is the single-phasing of a three-phase power system. Single-phasing may be caused by mechanical breakage of a conductor or by the blowing of a fuse. In most industrial plants, power-distribution circuits are built substantially and many of them are cabled, so that conductor breakage is a very rare occurrence. However, when a motor drives an isolated load, such as an irrigation pump in an arid area, it may be supplied over a comparatively weak transmission line, which may be subject to conductor breakage. A more common cause of single-phasing is the blowing of a fuse which protects a motor branch circuit, a feeder circuit on the secondary side of the transformers, or a feeder circuit supplying the primary side of the distribution transformers. When circuit breakers are used in the power-distribution system and in the motor-branch circuits, which open all three conductors simultaneously, single-phasing hardly ever occurs. However, since the builder of a control equipment has generally no knowledge of the details of the power system to which the controller may be connected, the possibility of single-phasing should be taken into consideration in selecting the motor protective devices.

When the branch circuit supplying a motor is single-phased, current flows through two of the conductors leading to the motor, but not through the third. A single-phase voltage is applied to two motor terminals. This single-phase voltage system can be considered as two voltage systems rotating in opposite directions, namely a positive-sequence system and a negative-sequence system of equal magnitude. As compared with the regular three-phase operation, the three stator phase windings do not carry the same current; hence they do not contribute evenly to the output of the motor. The following changes take place in the behavior of the motor.

When the motor is at standstill when the single-phasing occurs, it will not start upon being energized but will remain at standstill, drawing locked-rotor current. The locked-rotor impedance per motor phase does not differ significantly whether the motor is connected to a single-phase or a three-phase supply. Hence, the locked-rotor current drawn from the line during single-phase operation bears a ratio of $1.73/2 = 0.866$ to the locked-rotor current drawn during three-phase operation. Hence, interrupting the locked-rotor current of a single-phased motor is no problem to a line contactor. The heating in the motor is approximately equal to the heating of a three-phase motor drawing the same locked-rotor current. Although there is a somewhat increased hot-spot heating in the stator winding, because of changes in the spatial distribution of heat, this effect is not significant, at least not in the case of general-purpose motors. Tests and operating experience indicate that when thermal overload relays are so selected that they protect the three-phase motor under locked-rotor condition, these relays will also protect the single-phased motor under locked-rotor condition. Since the same current flows in two lines to the motor, two overload relays will provide adequate protection, since at least one relay will be in the circuit, irrespective of which phase is interrupted.

When a three-phase motor is single-phased while it is running, the motor will continue to run as a single-phase motor, provided the load torque does not exceed the breakdown torque of the motor. Its slip will be somewhat increased. If the motor is lightly loaded so that slip is not increased significantly, the current drawn from the line during single-phase operation is 1.73 times the current drawn during three-phase operation at the same load. However, when the motor operates closer to full load, it runs at higher slip, its power factor is reduced, increased losses due to negative-sequence current become more pronounced, and the current drawn from the line during single-phase operation becomes greater. It may run as high as 2.25 times the current drawn during three-phase operation. This current flows in two lines. Therefore, if two overload relays are used, at least one of these relays will see the current drawn by the motor irrespective of the phase which is open-circuited. Hence, two overload relays will pro-

MOTOR PROTECTION

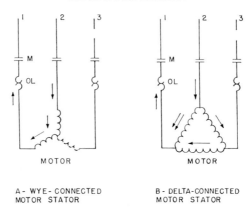

Fig. 15-6 Current distribution in wye- and delta-connected motor stator windings during single-phase operation.

vide adequate protection of the single-phased motor against a running overload.

Current drawn from the line which is seen by the overload relays may not determine the actual heating of the motor stator winding. There is a difference between motors with wye- and delta-connected stator windings, as illustrated in Fig. 15-6. When the stator is wye-connected, only two phases of the stator winding carry current, and that current is equal to the current drawn from the line. Hence the temperature rise of the operative windings is determined by the same current as is seen by the overload relay in the line.

In the case of a delta-connected stator winding all three phases carry current, but the current in one phase is approximately twice the current carried in the other two phases. Damage to the motor depends on the maximum temperature attained by the winding carrying the higher current. In contrast to three-phase operation, the current in the one winding is increased by a ratio of $2/1.73 = 1.15$ as compared with the increase in line current. For this reason it is customary in the United States to design motors built with delta-connected stator windings with a maximum current density in the stator conductors which is only 0.866 of the limiting current density used in the design of wye-connected motors, unless it is known that other means than thermal overload relays are provided for protecting the motor against single-phasing. This precaution assures that the temperature rise of the most highly loaded winding phase during single-phase operation is not higher than the temperature rise within an equivalent wye-con-

SINGLE-PHASING

nected stator winding. On the other hand, many foreign-built motors use the same limiting current density for both wye- and delta-connected motors. It is important to check this factor when controllers built in the United States are used with motors built in other countries, since standard thermal overload relays may not protect delta-connected foreign-built motors during single-phase operation.

It may be asked to what extent a single-phased three-phase motor is capable of keeping on running and carrying load. The maximum load which the motor can carry during single-phased conditions is limited by the breakdown torque the motor may develop. A single-phase motor is capable of absorbing a power input of approximately half the power input of a three-phase motor of equal size. The ratio of power output at the motor shaft is of the same order of magnitude. The breakdown torque of a three-phase motor during single-phase operation can be expected to be between one half and two thirds of the breakdown torque obtained during three-phase operation. The breakdown torque of NEMA design B motors, which constitute the bulk of industrial motors, is not less than twice, and for smaller high speed motors may even be as much as 2.75 times, rated motor torque. Hence the breakdown torque of a single-phased motor can be expected to be equal to or greater than rated full-load torque. Consequently, a motor which is running while single-phasing occurs, and which does not carry an overload, can be expected to continue running and carrying its load although the current it draws exceeds rated full-load current.

From the foregoing paragraph it can be concluded that overload relays which would protect the motor under three-phase running conditions are adequate to protect the single-phased motor against running overloads and stalling, provided the motor is single-phased by the interruption of one phase of the motor branch circuit.

A somewhat different condition exists when a feeder is single-phased by, for instance, the opening of a fuse on a power distribution switchboard. In that case, a number of motors may be connected to that feeder and the interrupted phase of these motors is interconnected between them. As long as any motors are running, they will act as phase converters; and, individually, each motor will "see" a three-phase power supply which is more or less balanced. If a number of motors is connected to the power system, and the ratings of these motors do not differ greatly, a fairly well-balanced three-phase operation is maintained. The degree of unbalance to which any individual motor is subjected will depend on its own load and the load of other motors on the system. As long as fairly well-balanced three-phase operation exists, two overload relays protect the motors.

MOTOR PROTECTION

With certain combinations of motor loads, however, it is entirely possible that the currents in the motor lines become severely unbalanced, so that one line carries a substantially larger current than the other lines. If that higher current flows in the line which is not protected by an overload relay, the overcurrent condition may not be noticed and damage to the motor may result. Figure 16-6 illustrates two load combinations which require three overload relays to protect against single-phasing of a feeder. *A* illustrates the condition that a three-phase motor and a single-phase motor operate on the same feeder. Phase 3 is open-circuited. Phase 2 of the three-phase motor will feed current not only to its own motor windings, but also, through phase 3, to the single-phase motor. Thus phase 2 of the three-phase motor carries a substantially higher current than phases 1 and 3. If no overload relay were provided in motor phase 2, that phase of the motor would be overloaded and exposed to potential damage. This combination of a three-phase and a single-phase motor does not occur regularly on industrial drives. However it does occur on air-conditioning equipments using a three-phase motor for driving a compressor and a single-phase motor for driving a fan.

Another example is shown in *B*. A large motor and a small motor, such as a 10- and a 1-horsepower motor, are connected in parallel. When one phase is open-circuited—in the example shown it is phase 3—the two motors act as phase converters to maintain phase 3. Depending on direction of rotation of the motor, either phase 1 or phase 2 will carry extra current to maintain phase 3. As far as the large motor is concerned, this extra current may not be sufficient to produce a pronounced unbalance in the three phase currents. However, the amount of this extra current may be a substantial percentage of the rated current of the small motor, so that if it flows in the phase which does not have an overload relay, damage to the motor may easily result. Therefore, three overload relays must be used to assure protection of the small motor. This combination of a large and a small motor is often found on industrial production machinery, on which a main motor and a small auxiliary motor may be connected to a common set of fuses.

If the primary of wye-wye or delta-delta connected distribution transformers feeding the low-voltage system is protected by fuses, an opening of one of the primary fuses will single-phase the low-voltage system in much the same way as the opening of a fuse on the low-voltage side of the distribution transformer. Whether or not the currents in the motor circuits are unbalanced to the degree that three

SINGLE-PHASING

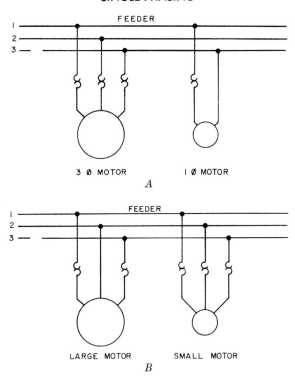

Fig. 16-6 Load combinations requiring three overload relays to protect against single-phasing of feeder. *A*. Combination of three-phase and single-phase motor. *B*. Combination of large and small three-phase motors.

overload relays are needed to protect the motors depends on the combination of motors operating on the secondary side of the transformers, their relative size, and their loading.

A peculiar condition exists when the secondary distribution system to which the three-phase motors are connected is fed from a wye-delta- or delta-wye-connected transformer bank. If one of the primary leads to such a transformer bank is open-circuited, unbalanced currents may flow in the secondary of the transformer, causing current to be excessive in one line only. The degree of unbalance will depend on the number of motors operating on the secondary side, the worst case being one motor operating on a transformer. This is a condition most frequently encountered on rural irrigation-pump installations, on which, for reasons of economy, fuses are generally used as primary transformer protection.

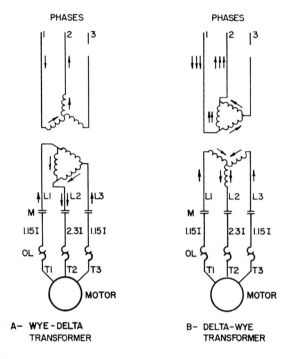

Fig. 17-6 Effect of loss of one primary phase on motor loads fed by wye-delta-connected and delta-wye-connected transformers. M = line contactor; OL = overload relays.

This condition is illustrated in Fig. 17-6. A shows a motor connected to a wye-delta transformer. Should one of the primary transformer phases be open-circuited for any reason—in the example it is phase 3—current would only flow in two legs of the transformer primary, as indicated by the arrows. Unbalanced current flows in the secondary circuit to which the motor is connected. If I represents the current which would flow in the motor under balanced conditions, $1.15I$ flows in two, and $2.30I$ flows in one secondary line. If a standard starter with two overload relays were used, the excessive current of $2.3I$ could flow in the line not protected by a relay. Therefore, to protect the motor fully against loss of any transformer primary phase, three overload relays must be provided in the starter.

A similar situation exists in motors fed by delta-wye-connected transformers, as shown in Fig. 17-6B. Unbalanced currents also flow when one primary phase is open-circuited, direction and relative magnitude of current flow being indicated by arrows. In this case, too, three overload relays are necessary to protect against single-phasing of the transformer primary under all conditions.

Since the use of wye-delta- and delta-wye-connected transformers is increasing, it has been suggested that the use of three overload relays on all three-phase a-c controllers be made compulsory. Such a step would increase the cost of all standard a-c controllers, and it could be justified only if single-phasing of transformer primaries and subsequent damage to motors were a frequent occurrence. The National Electrical Code recognizes this situation and it stipulates that inspection authorities may demand the use of three overload relays on motor starters connected to wye-delta and delta-wye transformer banks if experience at a given locality indicates that single-phasing of the transformer primaries is likely to occur in that locality.

Based on the existing provisions of the National Electrical Code, the industrial-control industry offers as standard starters with two overload relays. All manufacturers offer the addition of a third overload relay as a special form of starter. The question as to whether two or three overload relays should be the standard of the industry is being studied continually. Should changes in distribution system practices or future evidence of motor failures bring about a change in the provisions of the National Electrical Code, the control industry may change its standard in regard to number of overload relays at some future date.

PHASE-FAILURE PROTECTION

While overload relays suffice to protect continuously running motors against damage due to single-phasing in most instances, there are cases in which it is desired to disconnect the motor from the line upon the occurrence of a single-phasing condition. One such condition is the overheating of the rotor of certain squirrel-cage motors and of wound-rotor motors which cannot be protected by relays responding to stator current. Another condition is automatic reversing and plugging which is not possible when a single-phase voltage is applied to the motor terminals. Under these conditions, it is desirable to disconnect the motor from the line when single-phasing occurs. Phase-failure relays may then be used which close their contact when a balanced three-phase condition exists at the motor terminals, and which open their contact and cause the motor line contactor to be dropped out when single-phasing occurs.

282 MOTOR PROTECTION

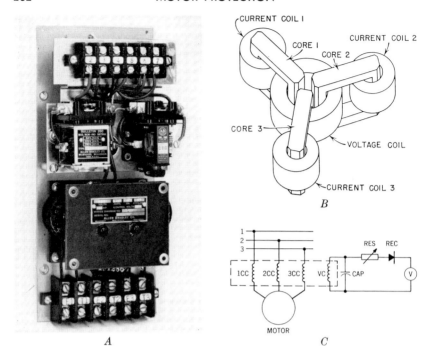

Fig. 18-6 Phase-failure relay. *A.* View of relay. *B.* Diagram of three-core current transformer. *C.* Basic relay circuit. (Courtesy Allen-Bradley Company.)

One form of phase-failure relay is illustrated in Fig. 18-6. *A* is a view of the relay which contains three essential elements. Above the bottom terminal board is a magnetic structure consisting of three current-measuring elements producing a control voltage which is used to sense a single-phasing condition. Above this unit are a potted network used for rectifying the voltage of the magnetic unit, which is then applied to a voltage relay which in turn opens or closes a contact connected in the coil circuit of the line contactor. The magnetic unit is illustrated schematically in sketch *B*. It consists essentially of three identical current transformers whose primary coils are connected in series with the motor to be protected. The three magnetic cores of square-loop magnetic material are so dimensioned that the minimum current of the lowest horsepower motor for a given size of coil will produce sufficient magnetomotive force for saturating the core at about 30 electrical degrees. Each primary coil surrounds one leg of each core, whereas the remaining legs are magnetically separated

PHASE-FAILURE PROTECTION

but surrounded by one common secondary voltage coil. Each flux induces a voltage in the common voltage coil, and the three voltages are totalized and applied to the coil of a voltage relay. The basic relay circuit is shown in C. Current coils $1CC$, $2CC$, and $3CC$ are connected in the motor lines acting on voltage coil VC. The voltage produced by VC is smoothed by a smoothing capacitor, adjusted by a variable resistor, and rectified. The rectified d-c voltage is applied to the coil of voltage relay V.

How the voltage output of the current transformer voltage coil can be used to discriminate between balanced and unbalanced voltage conditions is illustrated in Fig. 19-6. Diagram A illustrates conditions with balanced three-phase currents flowing to the motor. The solid sine wave labeled M represents the magnetomotive force due to current flowing in the current coils. The heavy solid line, which is a truncated sine wave represents the flux in each magnetic core, the flat portion indicating saturation. The broken sine wave represents the voltage which would be induced in the voltage coil if the core were not saturated. Actually, voltage is induced only during that portion of the sine wave during which the flux changes, and the shaded portion of the broken sine wave is the voltage E induced in the voltage coil. Two voltage pulses in opposite direction per cycle are induced for each phase, which add up to a voltage of basically three times line frequency. Diagram B illustrates the condition which exists when one line to the motor is interrupted so that only two lines carry current. The fluxes in these two coils are in exactly opposite phase positions, and consequently the voltage induced in the voltage coil is zero. Diagram C illustrates conditions which exist when the motor is supplied from a wye-delta-connected transformer and one of the primary transformer lines is open-circuited. One motor phase carries twice as much current as the other two phases. The combined fluxes produced by the current coils result in an output voltage having an average value approximately one third of that developed under three-phase symmetrical conditions. Diagram D illustrates the response of the voltage relay V. The average output voltage applied to the relay coil is plotted as a function of motor current for the three conditions that the motor operates on symmetrical three-phase voltages; that one transformer primary line is open-circuited; and that one line to the motor is open-circuited. The spread between symmetrical operation and single-phase operation is sufficiently wide so that drop-out voltage of the relay can be adjusted so as to discriminate with certainty between symmetrical and single-phased operation.

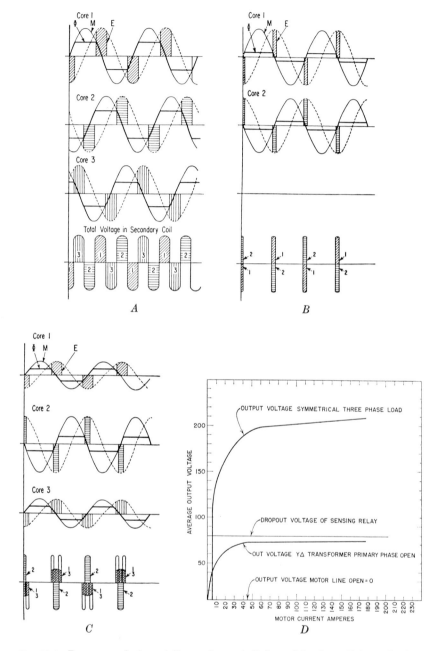

Fig. 19-6 Response of phase-failure relay. *A.* Balanced load condition. *B.* One line to motor open-circuited. *C.* One primary line of wye-delta transformer open-circuited. *D.* Voltage characteristic of relay coil. (Courtesy Allen-Bradley Company.)

PHASE-REVERSAL PROTECTION

Phase reversal or incorrect phase sequence in a three-phase circuit exists when connections for two of the three lines have been transposed. This condition may exist at the initial installation of a motor because of incorrect phasing of the motor lines, or it may be caused by reconnection of two lines somewhere in the power distribution system because of maintenance or repairs. As a result of phase reversal, the motor or motors in the circuit will rotate in the wrong direction. Damage may be caused to connected machinery, and personnel may be endangered. Phase reversal is particularly dangerous on equipment used for transporting people. Phase-reversal protection is a mandatory requirement of the National Electrical Code in connection with elevator control, and it is also called for in safety codes for machinery on which people may be transported.

Phase-reversal protection is provided by the use of a relay which closes its contact when phase sequence is correct and which opens its contact when phase sequence has been reversed. Such relays are used to shut down the motor in case of inadvertent phase reversal. Relays are available which respond either to motor current or to voltage on the motor branch circuit.

Figure 20-6 illustrates a current-responsive phase-reversal relay. Four magnets, the coils of which are excited by the current of two stator phases of the motor, set up a rotating flux which intersects a solid disk, pivoted in the center. Eddy currents are induced in the disk which develop a torque. This torque is made up of two components. A clockwise rotating torque is produced by the normal polyphase power. A counterclockwise rotating torque, due to the action of the pole shaders, is obtained on either polyphase or single-phase power. With balanced three-phase currents flowing, the clockwise torque is predominant and tends to turn the disk in a clockwise direction, but it is prevented from turning beyond a certain position by a projection in the disk which strikes a stop. If phase sequence reverses, the disk turns in a counterclockwise direction and the projection on the disk trips the contact open. If one phase to the motor is opened while the motor is standing still, or while the motor is carrying a load larger than approximately 50 per cent, the counterclockwise torque on the disk due to the pole shaders causes the disk to turn and open the relay contact. Thus the relay also provides phase-failure protection against starting the motor, and it will also stop a loaded motor. A lightly loaded motor would be permitted to continue its trip upon single-

MOTOR PROTECTION

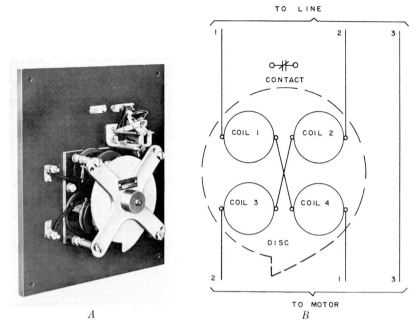

Fig. 20-6 Current-responsive phase-reversal relay. *A*. View of relay. *B*. Schematic diagram of relay connections.

phasing. After the relay contacts have been opened, they are reclosed by hand.

A voltage-responsive phase-reversal relay is illustrated in Fig. 21-6. This relay also consists of a set of four coils, a pivoted copper disk, and a snap-action contact unit containing one normally open and one normally closed contact. The coils are shunt coils, and they are connected to the three line phases ahead of the motor starter. When the motor branch circuit is energized with correct phase sequence, the disk rotates in a clockwise direction and actuates the contact unit. The normally open contact closes and permits energizing the motor starter. The normally closed contact opens; it may be used for signaling or some other auxiliary purpose. When phase sequence is incorrect, the disk rotates in a counterclockwise direction against a stop. This causes the contact unit to assume its normal position; that is, the normally open contact opens and drops out the motor line contactor. Thus phase-reversal protection is obtained with the motor running or at standstill.

PHASE-REVERSAL PROTECTION 287

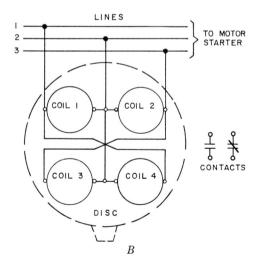

Fig. 21-6 Voltage-responsive phase-reversal relay. *A.* View of relay. *B.* Schematic diagram of relay connections. (Courtesy Allen-Bradley Company.)

This relay will also provide phase-failure protection to a limited degree. Should a phase open while the motor is at standstill, the relay opens its contact and drops out the line contactor. Thus the motor is prevented from starting on a single-phased line. But if the phase failure occurs while the motor is running, the voltage induced in the open-circuited motor winding will hold the relay in, thus permitting the motor to continue running single-phased. However, as soon as the motor is stopped, the relay will open its contact and prevent the motor from being restarted. An intermittent motor will thus be permitted to complete its trip before being shut down. This feature is valuable on equipments such as elevators. Upon loss of a phase, the elevator car will not stop suddenly between floors, but will continue to the next normal stop, and there it will be rendered inoperative.

BIBLIOGRAPHY

1. National Fire Protection Association, *National Electrical Code*, 1959 edition.
2. Underwriters' Laboratories, Inc., *Standard for Industrial Control Equipment*, May 1956.
3. C. W. Kuhn, "A-C Motor Protection," *Transactions AIEE*, May 1937.
4. G. A. Moffett, "Thermal Overload Protection for A-C Induction Motors," *Electrical Manufacturing*, July 1943.
5. D. B. Carson, "Overload Protection of Motors on Delta-Wye Supply," *General Electric Review*, June 1950.
6. G. W. Heumann, "Protective Controls for Low-Voltage Industrial Motors," *Electrical Manufacturing*, May 1951.
7. S. E. Hopferweiser, "The 'Ipsotherm' Motor Protection," *The Brown Boveri Review*, July/August 1951.
8. L. Matthias, "Thermal Overcurrent Relays for Motor Protection," *Electrical Manufacturing*, December 1951.
9. L. E. Daehler, "Overload Relays for Motor Protection," *Product Engineering*, August 1952.
10. S. E. Hopferweiser, "Investigations Concerning 'Ipsotherm' Motors Protection," *The Brown Boveri Review*, May 1954.
11. G. W. Heumann, "Motor Circuit Protection," *Machine Design*, June 1954.
12. J. Sheets, "Motor Overload Relays," *Product Engineering*, January 1955.
13. J. M. Bisbee, "Problems in the Application of Thermal Protection to Motors," *AIEE Paper* CP55-767.
14. G. W. Heumann, "Single-Phasing, What It Is, What to Do about It," *Power Engineering*, March 1957.
15. B. A. McDonald, "Protection of Three Phase Motors from Single Phasing," *IAEI News Bulletin*, March 1958.
16. V. G. Vaughan and R. M. Glidden, "Built-In Overheat Protection for Three-Phase Motors," *Electrical Manufacturing*, August 1958.
17. L. L. Gleason and W. A. Elmore, "Protection of 3-Phase Motors Against Single-Phase Operation," *AIEE Paper* 58-1018.

PROBLEMS 289

18. F. R. Karr, "Squirrel Cage Motor Characteristics Useful in Setting Protective Devices," *AIEE Paper* 59-13.
19. B. N. Gafford, W. C. Duesterhoeft and C. C. Mosher III, "Heating of Induction Motors on Unbalanced Voltages," *AIEE Paper* 59-29.
20. "Inherent Motor Overheat Protection Moves Inside the Field Coils," *Electrical Manufacturing*, November 1959.
21. J. K. Howell and J. J. Courtin, "Temperature Protection for Induction Motors . . . Today and Tomorrow," *Westinghouse Engineer*, November 1959.
22. T. F. Bellinger and R. A. Gerg, "Closer Overload Protection for Polyphase Motors," *Allis-Chalmers Electrical Review*, Second Quarter, 1960.

PROBLEMS

1. Define "overload protection" and list the most common causes of overload on a motor.

2. Describe the effect of overload on the motor and discuss specifically the following points:

(a) What are the generally accepted maximum temperature limits on an overloaded motor while running and while stalled?

(b) What is the effect of overloading on motor life?

(c) What is the permissible stall time of NEMA design B general-purpose squirrel-cage motors?

(d) What is the permissible stall time of wound-rotor motors?

3. Define "inherent protectors" and describe the principles on which their performance is based. Discuss in detail the following questions:

(a) What are the advantages and shortcomings of RTD's?

(b) What advantages have thermostats over RTD's, and what are their shortcomings?

(c) How do Klixon protectors differ from thermostats, and what are their advantages?

(d) Why are Klixon protectors not approved for general use, but only for use with specific motors?

(e) How are Klixon protectors connected in a dual-voltage motor circuit for wye- and delta-connected motors?

4. Answer the following questions in regard to thermal overload relay:

(a) What is the principle of a "replica" type relay and how does it function?

(b) Explain the difference between bimetal and solder-film relays.

(c) Explain the difference between resistance-heater and induction-type relays.

(d) What are the proper uses of hand-reset and automatically reset overload relays?

(e) How many overload relays are prescribed in the National Electrical Code for use with single-phase, two-phase, and three-phase motors, and in what conductors are they installed?

(f) Explain the effect of ambient temperature on relay performance.

(g) Discuss the reasons for tolerances in relay trip time and explain the significance of "tolerance band."

5. A manufacturer of thermal overload relays has laid out a line of relay heaters having rated ultimate trip current at 40 C, in accordance with the following table:

MOTOR PROTECTION

Number	Amperes	Number	Amperes	Number	Amperes
1	1.5	14	5.18	27	17.9
2	1.65	15	5.70	28	19.7
3	1.82	16	6.27	29	21.6
4	2.0	17	6.90	30	22.8
5	2.2	18	7.59	31	25.1
6	2.42	19	8.35	32	27.5
7	2.66	20	9.18	33	30.4
8	2.92	21	10.20	34	33.4
9	3.22	22	11.21	35	37.8
10	3.54	23	12.25	36	41.5
11	3.90	24	13.50	37	45.6
12	4.28	25	14.85	38	50.0
13	4.71	26	16.30	39	55.0

Select heaters which should be used for the following general-purpose, 40 C rise, three-phase induction motors, having a service factor of 1.15:

(a) 1 hp, 220 volts, 3.5 amp (b) 5 hp, 440 volts, 7.5 amp
(c) 5 hp, 220 volts, 15 amp (d) 7.5 hp, 220 volts, 22 amp
(e) 15 hp, 220 volts, 40 amp (f) 30 hp, 440 volts, 38 amp

6. Using the relay table of problem 5, select relay heaters for the motor as listed, but assuming that

(a) Motors are rated on a 50 C rise, no-service-factor basis.

(b) Relays are installed in a 40 C ambient room, and that temperature inside the controller may rise 15 C above room temperature.

7. A control manufacturer has published in his handbook the following heater selection table, listing the range of motor current for each heater, based on the use of 40 C rise general-purpose motors:

Number	Amperes	Number	Amperes
1	6.46– 7.18	14	18.5–20.3
2	7.19– 8.00	15	20.4–21.3
3	8.01– 8.55	16	21.4–24.6
4	8.56– 9.73	17	24.7–26.8
5	9.74–10.4	18	26.9–29.7
6	10.5 –11.2	19	29.8–33.6
7	11.3 –11.5	20	33.7–36.8
8	11.6 –12.2	21	36.9–41.0
9	12.3 –12.9	22	42 –47
10	13.0 –14.5	23	48 –55
11	14.6 –15.8	24	56 –62
12	15.9 –17.4	25	63 –66
13	17.5 –18.4	26	67 –73

Select from this table the heaters which should be used with motors rated, as follows:

(a) 25 hp continuous, 220 volts, 40 C rise, 64 amp
(b) 30 hp continuous, 440 volts, 50 C rise, 40 amp
(c) 10 hp, 1 hr, 440 volts, 55 C rise, 16 amp
(d) 15 hp, 30 min, 220 volts, 50 C rise, 44 amp
(e) 40 hp, 15 min, 440 volts, 50 C rise, 58 amp

8. A 30-horsepower, 440-volt, 38 amperes general-purpose, 40 C rise motor is used for driving a high-inertia load. Determine the type of relay and the maximum coil or heater current rating to be used under the following conditions:
(a) Relay installed in 30 C maximum ambient, accelerating time 18 seconds.
(b) Relay installed in 50 C maximum ambient, accelerating time 22 seconds.
(c) Relay installed in 40 C maximum ambient, accelerating time 7 seconds.
(d) Is there a second choice of relay which could be used under (c)?

9. What differences in thermal capabilities are there between standard general-purpose and hermetically sealed motors, and how are they protected against overloads?

10. For what motor applications are instantaneous overload relays used, and what is their setting?

11. Explain possible causes for single-phasing of three-phase motors.

12. Assume that a three-phase motor is single-phased by opening one line to the motor:
(a) What is the maximum torque the motor can develop?
(b) How do speed and line current change upon single-phasing?
(c) How much starting torque does the motor develop?
(d) What is the stalled-rotor current?
(e) How is motor heating affected?
(f) How many overload relays are needed to protect the motor?
(g) What is the difference between wye- and delta-connected motors, and how does this affect the choice of overload relays?
(h) Can all motors be fully protected by thermal overload relays?

13. Assuming that a motor feeder is single-phased by opening one line, determine how many overload relays are needed to protect motors under these conditions:
(a) One motor only is connected to the feeder.
(b) Several motors of substantially equal rating are connected to the feeder.
(c) One three-phase and one single-phase motor are connected to the feeder.
(d) One 25-horsepower and one 1½-horsepower, three-phase motor are connected to the feeder.

14. Assume that a wye-delta-connected transformer is used to supply a motor feeder:
(a) What is motor-current distribution in the three motor phases when one of the transformer primary lines is open-circuited?
(b) How many overload relays are needed to protect the motor, and for what reasons?
(c) How many overload relays are required by the National Electrical Code?

15. Discuss the principle of phase-failure protection, and name applications for which this protection is desirable.

16. (a) Discuss the principle of phase-reversal protection, and name applications for which this type of protection is mandatory.
(b) Explain the difference between current and voltage principle of phase-reversal relays.
(c) Do phase-reversal relays also provide phase-failure protection?

7 MOTOR BRANCH-CIRCUIT PROTECTION

The motor branch circuit is the connecting link between the power distribution system and the power utilization apparatus which, in most cases, is a motor. Design of the motor branch circuit is governed primarily by considerations of safety to the installed apparatus as well as safety to the machine operators. Improperly designed motor branch circuits may become a fire hazard. Therefore, the selection and installation of equipment comprising the motor branch circuit is governed to a large extent by the National Electrical Code.

In addition to the branch-circuit conductors, the branch circuit includes disconnect means, the motor controller, overload protection, and fault protection. Every motor controller includes overload protection. Disconnect means and fault protection may be included in the controller, or they may be provided by apparatus extraneous to the controller. Current rating of branch-circuit conductors for use with motors of given horsepower ratings is governed by tables in the National Electrical Code. Where motors are protected by inherent protectors, the branch-circuit conductors are protected by circuit breakers or fuses. When overload relays are used for protecting the motor, it can be assumed that the relays will also protect the branch-circuit conductors against overheating resulting from motor overload. Additional protection has to be included to safeguard the branch-circuit conductors as well as the connected equipment against the destructive effects of fault currents. This chapter will deal primarily with the protection of the branch circuit against short circuits.

PROTECTION AGAINST SHORT CIRCUITS

Short circuits may occur on any electric power system, and there are no economical means for preventing them entirely. Short circuits on motor branch circuits occur usually between the controller and the motor, within the motor, or at the controller load terminals. The latter types of faults can generally be traced to carelessness, and they

PROTECTION AGAINST SHORT CIRCUITS 293

occur during periods immediately following installation or servicing. They may be caused by wiring mistakes connecting two phases together inadvertently. They may also be caused by tools being left inside the controller or being dropped accidentally onto controller terminals. Faults within a motor are caused by deterioration of the winding insulation; they may result in short circuits between turns of the motor winding, or faults to ground. Short circuits within conduits are also caused by deterioration of the insulation of the branch-circuit conductors. Such deterioration may be due to mechanical injury when wires are pulled, contamination of the atmosphere, condensation of moisture within the conduit, or prolonged high temperature.

Since occasional faults are unavoidable, the branch-circuit conductors as well as the controller must be protected against the destructive forces of a short circuit. General-purpose industrial controllers are rated in amperes or horsepower. This rating is indicative of the normal load and of operating overloads, such as a stalled-rotor condition, which the control equipment is designed to handle. In an emergency, Class A controllers are capable of interrupting up to ten times their rated motor current. Such interrupting abilities are not at all adequate for handling the short-circuit currents that may flow under fault conditions in a modern industrial plant. In general, other devices are utilized to provide short-circuit protection.

The magnitude of the short-circuit current which may flow at any point in an industrial power system bears no relation to the rating of the motor and its associated control equipment installed on any branch circuit. The short-circuit current is determined solely by the characteristics of the power system. The short-circuit current which flows in a faulted system may be symmetrical or asymmetrical, as illustrated in Fig. 1-7. The symmetrical short-circuit current flowing at the location of the fault is equal to the system voltage divided by the sum of the impedances between the generator and the fault. During the first half cycle, an asymmetrical current of nearly twice the symmetrical short-circuit peak current may flow. The asymmetrical short-circuit current peak determines the maximum mechanical stress to which controller parts may be subjected.

Whether symmetrical or asymmetrical current flows in a conductor depends on the instant at which fault occurs. In a low-power-factor circuit, which a faulted circuit usually is, a symmetrical short-circuit current flows if the fault occurs when the voltage is near its crest value. A fully asymmetrical current flows if the fault occurs while the voltage is near zero. This asymmetrical current consists of the symmetrical

294 MOTOR BRANCH-CIRCUIT PROTECTION

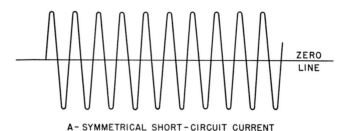

A- SYMMETRICAL SHORT-CIRCUIT CURRENT

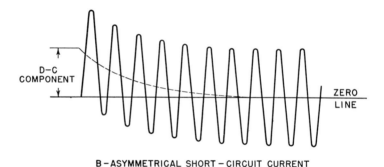

B - ASYMMETRICAL SHORT – CIRCUIT CURRENT

Fig. 1-7   Symmetrical and asymmetrical short-circuit current.

short-circuit current superimposed on (or offset by) a d-c component which decreases exponentially to practically zero within a few cycles. If the fault occurs at any other point of the voltage wave, the resultant short-circuit current is partially offset; that is, it contains a d-c component of reduced magnitude.

In a three-phase system, the fault may be a single-phase fault to ground with short-circuit current flowing only in one phase conductor. The fault may be phase to phase, in which case short-circuit current flows in two conductors. The most severe case is generally that of a three-phase short circuit in which all three phases are tied together and short-circuit current flows in all three phase conductors. The short circuit is then initiated at a different point of the voltage wave of each conductor. Therefore, the offset of the current varies in the three phases. Figure 2-7 is a typical oscillogram of a three-phase short-circuit current. It has been initiated so that phase A is fully offset, whereas phases B and C are partially offset. The most severe duty is imposed on the short-circuit-interrupting means when one phase is fully offset on a three-phase short circuit.

PROTECTION AGAINST SHORT CIRCUITS

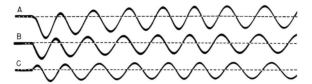

Fig. 2-7 Typical oscillogram of a three-phase short-circuit current. Phase A is fully offset; phases B and C are partially offset.

Devices which are intended to interrupt short circuits must be capable of interrupting the maximum current which may flow at the instant of current interruption. When the contacts of a circuit breaker part, current flow continues through an arc. Twice during each cycle, current passes through zero. Whether or not current flow is reestablished after a zero depends on the distance between contacts, that is the arc length, and the recovery voltage appearing across the contacts which may cause an arc to restrike. Since the short-circuit impedance of a power system is largely reactance, recovery voltage is usually at or near the peak value of the system voltage when current passes through zero. This is the most severe condition as far as probability of arc restriking is concerned.

When conducting tests to prove the short-circuit interrupting ability of circuit breakers or other current-interrupting devices, it is essential to consider the power factor at which the interrupting tests are conducted. Test codes included in recognized standards specify the maximum permissible power factor at which such tests may be conducted. In high-voltage branch circuits, the short-circuit power factor may be around 15 per cent. On low-voltage branch circuits, the short-circuit power factor may be as high as 45 or 50 per cent. The interrupting rating of a circuit-interrupting device at a given voltage may be expressed as the largest symmetrical rms current the device is capable of interrupting at that voltage. The manufacturer then establishes the rating so that the device is capable of interrupting a short-circuit current which includes whatever d-c component may be present at the instant of current interruption. With regard to circuit breakers, which find their largest use in distribution switchgear, it has become established practice to publish the interrupting rating as the actually available rms short-circuit current. Circuit breakers are applied on the basis of these published ratings. The application of a circuit breaker requires only knowledge of the available short-circuit current

296 MOTOR BRANCH-CIRCUIT PROTECTION

of the power system and of the location of the breaker within that system.

AVAILABLE SHORT-CIRCUIT CURRENT

Short-circuit studies of power systems may be rather time consuming. Such an effort is generally not warranted when the short-circuit current is to be determined which may flow in a motor branch circuit. Experience has shown that a greatly simplified method of calculating short-circuit currents obtains satisfactory results. A typical motor branch-circuit setup is illustrated schematically in Fig. 3-7. Modern industrial power systems are generally part of a large interconnected network in which the effect of the impedance of generators, step-up transformers, and transmission lines on a fault in the utilization branch circuits is negligible. Hence it can be assumed that the step-down transformer, feeding the branch circuit, is connected on its high side to an infinite bus, the voltage of which does not drop because of the short circuit on the low side of the transformer.

Any short circuit occurring within a controller is likely to incapacitate the controller, and back-up protection must be relied upon to clear the fault. Hence it is logical to expect the controller to clear a fault in the branch circuit between the controller and the motor terminals. The worst fault location to be considered, namely that resulting in the

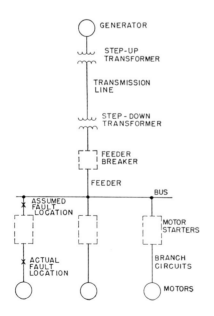

Fig. 3-7 Schematic diagram of a power system with motor branch circuits, indicating fault location on one branch circuit.

AVAILABLE SHORT-CIRCUIT CURRENT 297

highest short-circuit current, would be at the load terminals of the controller. Thus the fault impedance to be considered for calculating the short-circuit current actually flowing would be the sum of the impedances of the step-down transformer, the motor feeder, the controller, and the fault.

Fault impedance is a rather indeterminate quantity. It is assumed to be zero since, because of error in connections during installation, it may happen that two phases become bolted together. Impedance of the controller is a variable quantity. A large part of this impedance is the impedance of the overload relay heaters or overload relay coils, which varies with relay type and motor rating. It is difficult to appraise controller impedance. Therefore, it has become accepted practice in the control industry to neglect the controller impedance and to calculate the short-circuit current which would flow if the controller were short-circuited at its line terminals. This calculated current is the available short-circuit current at the point of controller installation. It is calculated by considering the reactance of the step-down transformer and the motor feeder. In a large plant which has its own power distribution system, a power system study has generally been made by the plant engineering department and the available short-circuit current is known at the principal points of motor installation. For a small plant connected to the distribution network of a public utility company, the available short-circuit current can generally be obtained from the power company.

Most short-circuit interrupting devices which are used at present in motor branch circuits are fast-acting, such as fuses, air circuit breakers, or contactors with high interrupting capacity. These devices clear the fault within less than three cycles after its incidence. During this short interval of time, induction motors and synchronous motors act as generators and contribute to the fault current. It is customary to consider this contribution by adding 100 per cent to the value of available short-circuit current as calculated with impedance between the infinite bus and the controller for 240-, 480-, and 600-volt systems, and 50 per cent for 208-volt systems.

When controllers include devices capable of interrupting short circuits, the interrupting capacity of the controller in indicated as the available short-circuit power which the controller is capable of interrupting. This means that the manufacturer warrants that the controller can be installed at any point in the power system in which the available short-circuit power does not exceed the interrupting rating of the controller. It is understood that the controller will be capable of interrupting any amount of short-circuit current which may actually flow on the load side of the controller. Performance tests are conducted by connecting the controller to a power supply capable, when short-circuited at the controller line terminals, of producing an available short-circuit current equal to the controller interrupting rating. During an interrupting test, the controller is short-circuited at its load terminals, and the controller interrupts the current which will actually flow with the added controller and fault impedance. The interrupting capacity of the controller is not necessarily identical with the interrupting ability of the device used for clearing the fault. In the interval of time during which short-circuit current flows, the controller is subjected to the thermal and mechanical stresses due to the short-circuit current. When subjected to a short circuit of full rated capacity, the controller must not be destroyed by the let-through energy prior to the instant at which short-circuit current flow ceases.

Available symmetrical rms short-circuit currents in low-voltage utilization networks frequently are of the order of tens of thousands of amperes. Quite often they go as high as 40,000 amperes, occasionally even up to 80,000 or 100,000 amperes. For low-voltage controllers, the interrupting capacity is presently expressed as rms asymmetrical amperes. To allow for the effect of the d-c component, the calculated symmetrical short-circuit current is multiplied by 1.25. There is a trend to express interrupting ability of low-voltage switchgear as symmetrical rms current. At a future date, low-voltage controllers may also be rated in terms of symmetrical rms current.

For high-voltage systems, it is customary to express short-circuit values in terms of available kva, which is simply the initial symmetrical rms short-circuit current times open-circuit phase-to-phase voltage times a phase factor (1.73 for three-phase systems). In 2400-volt and 4160-volt systems, short-circuit values as high as 150,000 kva and 250,000 kva, respectively, are frequently encountered. In some installations, it may be possible to obtain values as high as 350,000 kva. Since the interrupting capacity of high-voltage controllers is expressed as symmetrical short-circuit kva, the manufacturer assumes responsibility for evaluating the effect of the d-c component.

Contactors capable of interrupting fault currents are rarely found in low-voltage controllers. High-interrupting capacity contactors are more frequently found on high-voltage controllers. When contactor capacity is insufficient to interrupt the fault currents which may be encountered, other devices such as fuses and circuit breakers are used as fault-current interrupting means.

LOW-VOLTAGE FUSES

The National Electrical Code recognizes two types of devices for the protection of low-voltage branch circuits, namely fuses and circuit breakers. Fuses are among the oldest devices used for overcurrent protection. Their dimensions, ratings, and trip characteristics are standardized by NEMA for current ratings up to 600 amperes. Class H fuses—also known popularly as "NEC fuses," because they are built to dimensions which were formerly given in the National Electrical Code—usually have a zinc-base fusible conductor. Such fuses are tested by Underwriters' Laboratories, Incorporated, on a d-c circuit having an available current of 10,000 amperes. Considerable variations exist in the a-c interrupting ability of fuses of different current ratings and of different manufacture. While they have no stated a-c interrupting rating, various makes and sizes of Class H fuses have been tested and used successfully for motor branch-circuit protection on power systems with available short-circuit currents up to 20,000 amperes rms symmetrical, which corresponds roughly to 25,000 amperes three-phase average rms symmetrical. To permit disconnecting the branch circuit, motor circuit switches are generally used as disconnecting means in conjunction with fuses.

Under high short-circuit current, Class H fuses of the smaller sizes are "current-limiting." This behavior is illustrated in Fig. 4-7 for symmetrical and asymmetrical short-circuit currents. The dashed curves represent the available short-circuit current which would flow if no current interruption would take place. The solid curves represent the current which actually flows through the fuse and the controller protected by the fuse. During the initial rise of current in the first half cycle, sufficient heat is generated within the fuse to cause the fusible conductor to melt. An arcing period follows, during which the current drops to zero. Current is interrupted before it reaches its initial peak, during the first half cycle. Current-limiting action of the fuse reduces greatly the mechanical and thermal stresses which are imposed upon the controller by the short circuit. The current-limiting action, which is indicated by the ratio of peak let-through current actually flowing before interruption takes place to the initial peak current which would flow if no current interruption would take place, depends on the current rating of the fuse. Greater current-limiting

300 MOTOR BRANCH-CIRCUIT PROTECTION

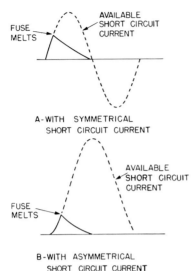

Fig. 4-7 Interruption of short-circuit current with current-limiting fuse.

action is obtained by fuses of low current rating than by fuses with a high current rating.

Besides the indeterminate interrupting capacity, which is not proven by tests, the principal disadvantage of Class H fuses is that the fusible conductors are subject to aging, and they are damaged by corrosion due to moisture. Zinc-base fuses age on the shelf, and they will eventually open-circuit with normal current flowing or with no current flowing at all. An open-circuited Class H fuse does not necessarily indicate that a fault has occurred on the branch circuit. These disadvantages have been overcome by the more recent Class J and Class L fuses, which are designed and tested to obtain a definite current-limiting action and a stated interrupting rating. Class L fuses are available in ratings up to 600 amperes. They have the same dimensions as Class H fuses, and they fit standard NEC fuse clips. Special "reject-type" fuse clips are available which permit only the use of Class L fuses but not of Class H fuses. Such clips may be used when it is desired to prevent the substitution of ordinary Class H fuses. Class J fuses can be obtained in current ratings up to 6000 amperes. Their dimensions differ from the NEC dimensions, and they require special fittings to hold them.

In outward appearance, high-interrupting capacity fuses Class L and Class J do not differ greatly from ordinary Class H fuses, as illustrated in Fig. 5-7. However, their internal construction is quite different.

LOW-VOLTAGE FUSES 301

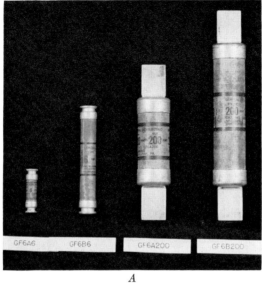

A

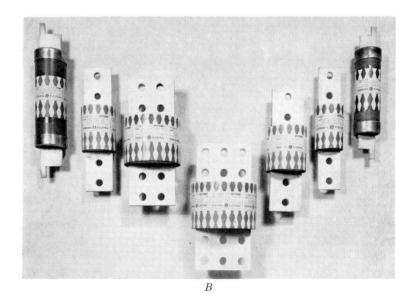

B

Fig. 5-7 High-interrupting capacity, current-limiting, low-voltage fuses. *A*. Class L, 6 and 200 amperes, 250 and 600 volts. *B*. Class J, 300 through 4000 amperes, 600 volts.

They contain fusible wires or ribbons made of silver, imbedded in a filling of quartz sand, and the fuse barrel consists of impregnated high-strength insulating tubing. The fusible links of such fuses are not subject to aging and corroding. Their shelf life is indefinite, and nuisance blowing of properly selected fuses is nonexistent. If a high-interrupting capacity fuse blows, it is a certain indication that the branch circuit has been faulted.

Class J and Class L fuses are available which have been proven by tests to interrupt successfully available short-circuit currents of 200,000 amperes symmetrical. Such fuses are capable of interrupting any short-circuit current which may occur on the largest present-day motor branch circuits.

LOW-VOLTAGE CIRCUIT BREAKERS

The other type of device recognized by the National Electrical Code for the protection of motor branch circuits are air circuit breakers with time-limit overcurrent trips. Such breakers have been described in Part I, Chapter 3. For large motors or as feeder protection of a group of motors, switchgear-type air circuit breakers may be used. Such circuit breakers can be obtained for manual or magnetic operation, and they can be equipped with inverse-time overcurrent trip elements. For branch circuits serving smaller and medium-size motors, molded-case circuit breakers are generally used. These breakers are mostly manually operated, and the standard forms are equipped with combination thermal and magnetic overcurrent trips.

Since the thermal trip units include bimetal elements, their tripping characteristic is influenced by the ambient temperature in which the circuit breaker operates. Molded-case circuit breaker standards are based on the assumption of an ambient temperature of 25 C. Few breakers used on branch circuits actually operate in such low ambient. For control equipment, the ambient temperature, that is the temperature of the room in which control equipment is installed, is assumed to be 40 C. When control equipment is enclosed, the temperature inside the enclosure is likely to rise higher than 40 C, and there are industrial locations in which temperatures inside enclosures may be as high as 60 C. When a molded-case breaker, with thermal-magnetic trip is used in a high-ambient location, and if the motor with which it is used drives a high-inertia load, the thermal trips may open the circuit breaker in a shorter time than the published tripping curve indicates. This may result in nuisance tripping before the motor has completely accelerated. The effect of ambient temperature on the

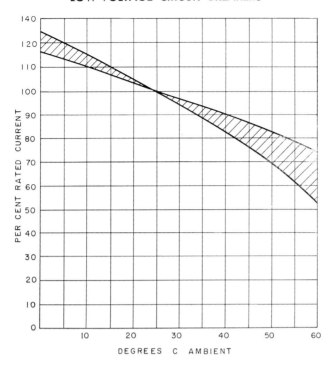

Fig. 6-7 Effect of ambient temperature on molded-case air circuit breaker thermal trips.

breaker thermal trip characteristic is shown in Fig. 6-7. The shaded band indicates the range of conversion factors which indicate the breaker performance at ambients differing from 25 C. Temperature-conversion factors for specific types of breakers would have to be obtained from the breaker manufacturer. For example, when operating in a 50 C ambient, a 100-ampere breaker may perform according to its published tripping curve for an 80-ampere rating when used on a motor load. Breakers used with high-inertia loads in high ambient temperatures must be carefully selected, or they should be equipped with thermal trips specially calibrated for performance in higher ambient temperatures.

The instantaneous trips, being magnetic, are not affected by ambient temperature. When applying breakers with specially calibrated thermal trips in high-ambient locations, the proper relationship between thermal and magnetic trip units must be maintained to avoid instantaneous tripping on too high a percentage of rated current of the motor connected to the circuit. Breakers having only instantaneous trip elements set at approximately twelve times motor full-load current have been used successfully. The use of such breakers requires that the starter be capable of withstanding overload and fault currents up to the trip setting of the breaker.

COORDINATION OF FUSES AND BREAKERS WITH MOTORS

Faults may be between phases or from phases to ground. Hence, fault current may only flow in one conductor of a motor branch circuit. For this reason, the National Electrical Code prescribes that fuses or circuit breaker trips be provided in each ungrounded line to the motor. Three fuses or three overcurrent trips must be used for the protection of three-phase motor branch circuits.

Selection of the current rating of fuses and circuit breakers is governed by the ability of the controller to withstand the effects of the let-through current which flows before the short circuit is interrupted. The most vulnerable parts of a motor controller are the heaters or coils of the overload relays. Breakers and fuses are so selected that the thermal overload relays will not be destroyed and initiate an internal short circuit in the controller on account of fusing due to the thermal effect of the let-through short-circuit current. Based on exhaustive studies and operating experience, the National Electrical Code has established maximum current ratings of fuses and circuit breakers in terms of motor full-load current. Table 9 indicates the maximum rating of fuses and circuit breakers which may be used with squirrel-cage motors marked with a code letter, that is to say with motors whose stalled-rotor current is known. Corresponding current values are given in Table 10 for use with wound-rotor motors and squirrel-cage motors which are not marked with a code letter. Fuses and circuit breakers selected in accordance with these tables will permit a motor to be started under average starting conditions, in other words, fuses will not blow and circuit breakers will not trip before the motor has been fully started. Should it be found that the motor cannot be started with fuses and breakers so selected, the next larger size fuse or breaker trip rating could be used; but under no circumstances must their rating exceed four times motor full-load current.

Special attention must be given to the selection of breakers and fuses for use with part-winding starters. In Chapter 2 (Fig. 32-2 and page 57) it is explained that two sets of overload relays are provided for protecting the motor during starting and running. Each set of

Table 9. Maximum Rating of Motor Branch-Circuit Protective Devices for Motors MARKED with a Code Letter

Type of Motor	Per Cent of Full-Load Current	
	Fuse rating	Circuit-breaker setting, time-limit type
All a-c single-phase and polyphase squirrel-cage and synchronous motors (full-voltage, resistor, or reactor starting)		
Code letter A	150	150
Code letters B to E	250	200
Code letters F to V	300	250
All a-c squirrel-cage and synchronous motors with autotransformer starting		
Code letter A	150	150
Code letters B to E	200	200
Code letters F to V	250	200

relays is selected for one half motor full-load current. To maintain the four-times relationship to overload relay current, circuit breakers and fuses used on part-winding starters should have a current rating of not more than two times motor full-load current.

Proper correlation of thermal overload relay and a molded-case circuit breaker is indicated in Fig. 7-7. The diagram has been drawn under the assumption that a general-purpose 40 C rise motor has been used. The ultimate trip current of the overload relay is assumed to be 125 per cent of motor full-load current, and the trip-unit rating of the circuit breaker has been selected as 250 per cent of motor full-load current. The trip characteristics of the relay and the breaker are drawn as lines in order to simplify the diagram. A complete correlation study should include the tolerance band of the tripping characteristic of the two devices. The first criterion for proper correlation is the protection against operating overloads. When a motor stalls at approximately six times full-load current, the thermal overload relay should trip first so that the motor is removed from the line by the contactor without tripping the breaker. The second requirement which has to be met is the protection of the overload relay. Thermal

Table 10. Maximum Rating of Motor Branch-Circuit Protective Devices for Motors NOT MARKED with a Code Letter

	Per Cent of Full-Load Current	
Type of Motor	Fuse rating	Circuit-breaker setting, time-limit type
Single-phase, all types	300	250
Squirrel-cage and synchronous (full-voltage, resistor, or reactor starting)	300	250
Squirrel-cage and synchronous (autotransformer starting)		
Not more than 30 amp	250	200
More than 30 amp	200	200
High-reactance squirrel-cage		
Not more than 30 amp	250	250
More than 30 amp	200	200
Wound-rotor	150	150

overload relays of the heater type are considered to be self-protecting up to between ten and twelve times rated motor current. This means that with current of less than that limiting value flowing the overload relay trips and removes the motor from the line before it has been damaged by overheating. If the relay is subjected to current greater than this limit, there is danger that the overload relay is damaged by overheating before it trips the line contactor. At currents equal to or larger than the relay self-protecting limit, the circuit breaker should open first, thereby preventing damage to the overload relay. The maximum interrupting ability of the line contactor should also be taken into consideration, and the circuit breaker should open first on any current larger than the interrupting ability of the line contactor. In Fig. 7-7, the thermal overload relay and the breaker trip curves intersect at approximately eight times motor current, which means that the overload relay as well as the line contactor are protected against possible damage. Diagrams to study the correlation between overload relays and fuses are made in a similar manner, using time-current curves from the fuse manufacturer.

Induction-type overload relays are self-protecting up to larger cur-

COORDINATION OF FUSES AND BREAKERS WITH MOTORS 307

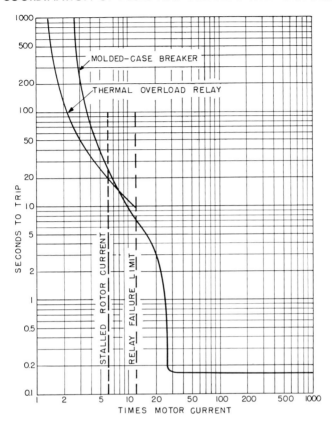

Fig. 7-7 Coordination of thermal overload relay and thermal-magnetic circuit breaker.

rents. Because of the higher thermal withstand ability of the coil, induction relays can be assumed to be self-protecting up to a current about twenty-five times motor full-load current. If more accurate data are needed, they should be obtained from the relay manufacturer.

When instantaneous-trip circuit breakers are used for the protection of motor branch circuits, serving squirrel-cage motors, the instantaneous trips must be adjusted for a sufficiently high current to permit the motor to accelerate to full speed without nuisance tripping of the breaker. Since the instantaneous trip unit of the breakers operates within one cycle, it does not suffice to consider the steady-state symmetrical starting current as a basis for overload trip setting. The breaker must be so set as not to trip during the initial first-cycle peak of the starting current. Because of saturation in the motor iron, the initial symmetrical peak current is likely to be higher than six times rated motor current. Furthermore, the initial inrush current may be fully offset, although the d-c component as well as the effect of saturation decay very rapidly. An instantaneous trip setting of twelve to fifteen times motor full-load current may be required to take care of all possible variations in motor starting currents.

Before instantaneous-trip circuit breakers are used for branch-circuit protection, a careful check must be made to ascertain that overload relays and line contactors will not be damaged by high-impedance fault currents just below the trip setting of the breaker. Because the use of instantaneous-trip circuit breakers for branch-circuit protection requires special study and special selection of components, their use is not recognized by the National Electrical Code. For any installation subject to inspection under National Electrical Code rules, a waiver should be obtained before instantaneous-trip breakers are supplied. The use of such breakers is not recommended for reversing plugging drives, since the high initial current peak upon plugging would require an undesirably high circuit breaker trip setting.

CLASS C CONTROLLERS

The National Electrical Code does not specify where fuses or circuit breakers are to be arranged in the branch circuit as long as they are connected ahead of the motor and the starter. They may be part of the motor controller, or they may be supplied as separate items, to be assembled and wired by the user. When they include fused motor circuit switches or circuit breakers, starters are designated as "combination starters." Typical combination starters are illustrated in Fig. 9-2. In addition to their usual function of motor protection, such starters also include the function of branch-circuit disconnection and branch-circuit protection.

Controllers having a high interrupting rating, beyond the interrupting ability of ten times rated motor current, are designated as "Class C controllers." Some special controllers are built with contactors having an inherently high interrupting ability, and such contactors can be used for interrupting short-circuit currents. However, most Class C controllers utilize fuses or circuit breakers for interrupting fault currents, and ordinary contactors with ten times motor-current interrupting ability for starting and stopping the motor. Class C controllers have stated interrupting ratings. Most frequently used ratings are 15,000 and 25,000 amperes with molded-case circuit breakers,

CLASS C CONTROLLERS

25,000 amperes with Class H (or NEC) fuses, and 50,000 amperes with Class J or L high-interrupting capacity fuses.

Class C controllers offer the advantage to the user that the manufacturer assumes the responsibility for proper coordination of all the components in the controller. It must be understood that the interrupting rating of a Class C controller is not only determined by the interrupting ability of the device used for short-circuit interruption. The current-carrying parts of the starter, the line contactor, overload relays, buses and interconnecting power wiring must be capable of withstanding the let-through energy which flows through the starter prior to interruption of the short-circuit current. Not only must the current-carrying parts have sufficient thermal ability; they must also be capable of withstanding the mechanical stresses due to peak let-through current. Generally speaking, the interrupting rating of a combination controller can be controlled and proven by test much more reliably than the adequacy of a combination of unrelated individual components on a branch circuit.

The National Electrical Code recognizes circuit breakers and fuses as adequate short-circuit protective means, and no indication is given of any preference for either. However, practical considerations may influence the choice between breakers and fuses. With regard to let-through energy, which determines the thermal and magnetic stresses on the starter, current-limiting fuses greatly reduce these stresses. This feature is an important consideration, particularly in connection with smaller starters, say up to and including size 2. Circuit breakers of the types generally used for low-voltage branch-circuit protection cannot be obtained with trip ratings of less than 15 amperes, which may not provide adequate protection for overload-relay heaters of small motors say 3 horsepower or less. For this reason, many users prefer fuses for the protection of small motor branch circuits. With larger-size starters, the current-limiting effect of fuses becomes less pronounced, and the larger starters have comparatively greater thermal and mechanical strength than the smaller sizes. Hence circuit breakers are often preferred for use with starters size 3 and larger.

From an operating point of view, circuit breakers have the advantage that, when they have been tripped, they can be reclosed by a machine operator. When fuses are used, it is necessary to open the controller enclosure for replacement of fuses. In many plants, safety regulations prevent floor operators from opening enclosures, and an electrician has to be called, with resultant delay in restarting the machine, in the event that an electrician is not immediately available. For this reason many plant operators prefer circuit breakers, feeling that their use reduces possible down time.

As long as Class H fuses were the only type available, nuisance blowing of fuses occurred and production was interrupted for no reason. Class J and L fuses, which use silver as material for the fusible links, have indefinite shelf life, and they blow only when there is an actual fault. In that case, it is preferable to have an electrician come and check the branch circuit before replacing fuses. Indiscriminate reclosing of a circuit breaker by a machine operator upon a short circuit may create a hazard.

BREAKER-FUSE COMBINATION

Severe faults with practically zero fault impedance, which would require the full interrupting ability of the controller, occur only rarely on a branch circuit. Many faults are accompanied by considerably less current flow than that corresponding to the interrupting rating of the controller. With the advent of current-limiting high-interrupting capacity fuses, which have a predictable current-limiting action and a proven interrupting ability, molded-case circuit breakers and fuses may be used in combination, and they are suitable for operation on a power system having an available short-circuit current corresponding to the interrupting rating of the fuse. Low short-circuit currents, within the interrupting ability of the circuit breaker, are cleared by the breaker without melting the fuse. After the fault has been cleared, the breaker may be reclosed and no fuse replacement is necessary. If a fault occurs with current flow beyond the interrupting ability of the breaker, the fuse takes over and clears the fault. When a fuse of proper current rating is selected, its current-limiting action and short clearing time prevent damage to the breaker. This combination offers the operating advantage that fuse replacement is only necessary in the comparatively infrequent cases of high-current faults. Satisfactory operating results and proper selectivity can be obtained with standard breakers and standard Class J and L fuses.

So-called "current-limiting circuit breakers" have been developed which combine a molded-case circuit breaker with a set of current-limiting, high-interrupting capacity fuses in a common assembly. Such breakers are available as Cordon * circuit breakers built by the ITE Circuit Breaker Company. Similar breakers are built by the Westinghouse Electric Corporation under the name of Tri-Pac.** Figure 8-7

* Registered trade mark of the ITE Circuit Breaker Company.
** Registered trade mark of the Westinghouse Electric Corporation.

BREAKER-FUSE COMBINATION

Fig. 8-7 Cordon current-limiting circuit breaker. *A*. External view of assembled breaker. *B*. Exploded view of breaker with cover removed, showing assembly of current-limiting fuses. (Courtesy ITE Circuit Breaker Company.)

A

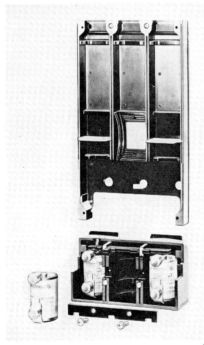
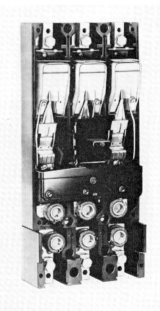

B

312 MOTOR BRANCH-CIRCUIT PROTECTION

illustrates the construction of a Cordon circuit breaker. A is a view of the complete breaker with cover in place. In appearance, it resembles a standard molded-case circuit breaker, except that a fuse assembly is added at the bottom of the breaker. B is an open view showing the cover removed and exposing the inside of the breaker. At the bottom is the plug-in unit containing three current-limiting, high-interrupting capacity fuses, which are plugged into receptacles in the lower part of the circuit-breaker base. A mechanical linkage interlocks the fuse unit with the circuit-breaker trip unit so that the fuse unit can only be removed with the circuit breaker tripped. To avoid single-phasing of the branch circuit, a spring-loaded plunger in each fuse trips the circuit breaker whenever one fuse opens. This arrangement assures that all three phases of the branch circuit are interrupted when any one fuse blows. The breaker cannot be reclosed unless the blown fuse

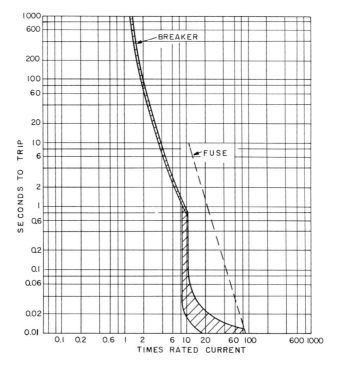

Fig. 9-7 Trip characteristic of 100-ampere Cordon circuit breaker. (Courtesy ITE Circuit Breaker Company.)

has been replaced first, thus assuring that the breaker cannot be reclosed upon a single-phased branch circuit.

A typical trip characteristic of a 100-ampere Cordon circuit breaker is given in Fig. 9-7. The solid curves represent the trip characteristic of the molded-case-breaker unit when operating at normal 40 C ambient, and the shaded area represents the tolerance band. The dashed curve is the trip characteristic of the fuse. For short-circuit currents up to approximately 100 times rated breaker current, that is 10,000 amperes, the breaker trips before the fuse melts. At currents beyond that value the fuse opens within the first half cycle after the incidence of the fault. Such circuit breakers have been used on combination controllers having an interrupting rating of 50,000 amperes and more.

MOTOR CONTROL CENTERS

In Part I, motor control centers have been discussed as a coordinated mechanical assembly of motor starters for a plant or a section of a larger plant. In addition to the convenience and orderly mounting arrangement of the starters, motor control centers perform the important function of centralizing the branch-circuit protection for a manufacturing area. In a motor control center, the design of starters, buses, and interconnecting power connections is so correlated that it takes into account the hazards which short-circuit performance involves.

In a motor control center, each starter compartment serves a specific power utilization equipment, and there is no general power distribution to outside subscribers. To obtain a reasonable economic balance purchasers generally do not expect the same degree of continuity of service from a motor control center which they would expect from a power distribution switchboard. When a branch circuit is faulted, some maintenance of components may be required before the branch circuit can be placed in service again. Contactor contacts may freeze, overload relay heaters may be damaged, circuit breakers may be out of calibration when a short circuit of full rated capacity occurs. In some cases, components may even need replacement. In any event, however, the effect of the short circuit will be localized to the compartment of the starter on whose branch circuit the fault occurred.

Motor control centers should be so designed that faults are not communicated from one compartment to another. If a short circuit occurs in one starter compartment, for instance because of carelessness

MOTOR BRANCH-CIRCUIT PROTECTION

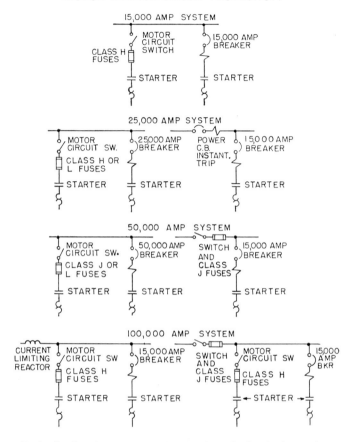

Fig. 10-7 Basic circuits of motor control center branch circuits for various amounts of available short-circuit current.

of an electrician, this compartment may burn out. However, with good control center designs, the fault will not be communicated to the bus structure and cause extensive damage to the rest of the equipment. Motor control centers are rated in available three-phase average asymmetrical short-circuit current at its incoming line terminals. In Fig. 10-7, various methods are indicated by which certain interrupting ratings can be obtained on the complete control center. These ratings are established on the assumption that faults will occur on the load side of the individual starters. Bus faults within the control center have to be cleared by the feeder breaker serving the control center.

The lowest standard rating is 15,000 amperes. Starters are directly

MOTOR CONTROL CENTERS 315

connected to the bus. Short-circuit protection is obtained either with a motor circuit switch and Class H (NEC) fuses or a molded-case circuit breaker having an interrupting rating of 15,000 amperes.

A 25,000-ampere control center rating is obtained when motor circuit switches are used with Class H (NEC) or L (high-interrupting capacity) fuses or circuit breakers having an interrupting rating of 25,000 amperes. It is also possible to use 15,000-ampere-rated molded-case circuit breakers on the individual motor starters, provided the main bus of the control center is protected by a power circuit breaker with instantaneous overcurrent trips. These trips should be set sufficiently low so that, on a short circuit beyond the interrupting ability of the individual motor starters, the power circuit breaker trips to clear the fault and prevents destruction of the starters on the individual branch circuits, even though starter components may be damaged. On the other hand, the trip units of the power circuit breaker must be set sufficiently high so that short circuits within the capacity of the individual starter breakers are cleared by these breakers and only the faulted branch circuit is shut down, without taking power off the whole control center.

A 50,000-ampere rating can be realized with the use of motor circuit switches with Class J or L (high-interrupting capacity) fuses on each motor starter, or the use of individual circuit breakers having a 50,000-ampere rating. It is also possible to use 15,000-ampere-rated circuit breakers on the individual motor starters and a set of Class J current-limiting fuses in the main bus. Fuse size has to be carefully coordinated with breaker let-through ability, and so selected that the let-through energy of the fuses is sufficiently low to prevent destruction of the breakers when any branch circuit of the motor control center is subjected to a full-intensity short circuit. The fuses selected must also be sufficiently large so that fault currents within the rating of the circuit breakers will not cause them to blow.

There are two methods of obtaining a 100,000-ampere rating. One is to use current-limiting reactors in the main bus. These reactors are so selected that they reduce the maximum available short-circuit current to a value within the capabilities of the combination starters on the individual branch circuits. As an example, the current-limiting reactors may reduce the maximum available short-circuit current to 15,000 amperes, thereby permitting the use of motor circuit switches with Class H (NEC) fuses or 15,000-ampere-rated circuit breakers. It must be recognized that the current-limiting reactors introduce an additional impedance in series with the feeder serving the motor control center, which results in additional voltage drop on the control center bus. This additional voltage drop must be taken into consideration when motors connected to the control center are highly loaded, since the maximum torque which can be obtained from the motor is reduced. Likewise, it must be ascertained that the additional drop will not interfere with the successful closing of magnetic devices forming part of the motor controllers. A 100,000-ampere rating can also be realized without the use of current-limiting reactors by installing a set of Class J current-limiting fuses in the main bus, and the same considerations govern as have been described in the preceding paragraph in connection with control centers suitable for use on 50,000-ampere systems.

When a motor control center is installed adjacent to a unit substation, the available short-circuit current is determined by the rating of the transformer to which the control center is connected. In Table 11,

Table 11. Largest Transformer Rating to which Motor Control Center Can Be Connected with No Feeder Impedance

Interrupting Rating (amperes)	Transformer kva at		
	240 volts	480 volts	600 volts
15,000	200	400	500
25,000	300	750	900
50,000	600	1500	1800
100,000	1200	3000	3600

the maximum transformer ratings are listed to which motor control centers of a given interrupting rating can be connected. The transformer ratings represent average conditions based on customary transformer impedances, including contribution to the short-circuit current by induction and synchronous motors which may be connected to the control center and which may be running at the time the fault occurs. These transformer ratings should not be exceeded unless a system study which considers in detail the power system and transformer impedances, as well as the impedance of the feeder between transformer and control center, indicates that the available short-circuit current is within the capabilities of the control center.

CLASS E—HIGH-VOLTAGE CONTROLLERS

In bygone years, when industrial power systems were small and available short-circuit currents amounted to only a few thousand amperes, Class A controllers, having an interrupting ability of ten times rated motor current, were used extensively on high-voltage branch circuits. Slow-acting oil circuit breakers at the substation were relied upon as back-up protection against short circuits. Blowups of contactors occurred quite frequently. With present interconnected power systems, it is acknowledged that such practices are no longer adequate. Today, NEMA does not recognize as standard any Class A high-voltage controllers. Because of the severe damage which can be done by a short circuit on a modern 2400-volt or 4160-volt power system, high-interrupting capacity controllers should be used exclusively for high-voltage motors.

NEMA has established Class E as a designation for high-voltage, high-interrupting capacity a-c controllers. This category includes two kinds of controllers. Class E1 controllers are those on which the main contacts are used both for starting and stopping the motor, and for interrupting short-circuit currents. Class E2 controllers utilize a contactor for starting and stopping the motor, and fuses for short-circuit interrupting. The interrupting time of contactors or breakers used on Class E1 controllers may vary considerably between sizes and types. On Class E2 controllers, current-limiting or power fuses may be employed. Depending on the character of the power system to which the controller is connected, the magnitude and time constant of the d-c component of the short-circuit current may vary. Hence controller rating based on total interrupted short-circuit kva would make controller selection a rather involved procedure.

To simplify controller application, present industry standards base the equipment rating and controller application on available symmetrical short-circuit current of the power system at the point of controller installation. Consequently, Class E controllers are rated in maximum available symmetrical short-circuit kva at the controller line terminals. This procedure simplifies the listing of high-voltage controllers of various manufacturers, since all controllers suitable for use on a certain motor branch circuit have the same rating. Any differences in actual interrupting duty, expressed as interrupted total kva, are of no concern to the user. They are taken into consideration by the controller manufacturer in the design of the controller parts. The testing procedure for verification of the symmetrical kva interrupting rating is

MOTOR BRANCH-CIRCUIT PROTECTION

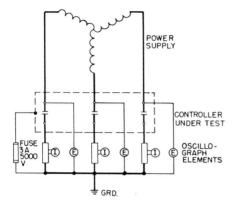

Fig. 11-7 Standard NEMA test circuit for Class E controllers.

so devised as to subject the controller to a total kva interrupting duty, which is representative of the duty to which the controller will be subjected at the installation.

In accordance with NEMA standards for Class E controllers, interrupting rating is verified by a test circuit in accordance with Fig. 11-7. First, the test circuit is calibrated without the controller in the circuit. With the controller short-circuited at its line terminals, the test circuit must be capable of producing an rms symmetrical short-circuit current which is not less than the symmetrical rms current on which the short-circuit kva rating of the controller is based. The symmetrical current flowing during the three-phase test is taken as the average of the currents in the three phases.

In order to consider the effect of d-c offset in the three-phase current (see Fig. 2-7), NEMA standards specify that the test circuit must be capable of producing in one of the three phases a total rms current which, at various times after initiation of the short circuit, bears a ratio to the rms symmetrical current on which controller rating is based, as given in Table 12. This ratio is represented graphically in Fig. 12-7. A further requirement of the testing procedure is that the power factor of the test circuit shall not exceed 15 per cent lagging, to insure that current zero occurs in the vicinity of crest system voltage, thus insuring high recovery voltage. Also, the open-circuit line-to-line voltage must not be less than the voltage on which the interrupting rating is based.

To permit proper adjustment of the desired test current, resistors, reactors, and transformers may be included in the test circuit in addition to the generating system. When setting up a controller for testing,

CLASS E—HIGH-VOLTAGE CONTROLLERS

Table 12. Ratio of RMS Total Current to RMS Symmetrical Current for Class E Controller Tests

Time after Initiation of Short Circuit in Cycles (60-cycle basis)	Ratio of Total RMS Current in the Phase with Maximum D-C Component to RMS Symmetrical Current Corresponding to Interrupting Rating in KVA
½	1.6
1	1.4
2	1.2
3	1.1
4 or more	1.0

the leads to the controller are made as short as practicable so as to keep capacitance to ground at the controller terminals as small as possible, and thereby obtain a high rate of rise of the recovery voltage. No capacitance may be added to the test circuit. A synchronous switch may be used for initiating the calibration shots. The synchronous switch is so adjusted that maximum offset is obtained in one phase. When no synchronous switching facilities are available, several random shots may be taken in order to obtain a shot with maximum offset in

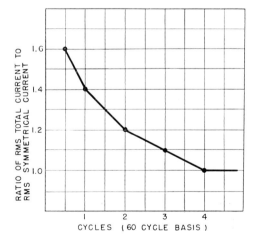

Fig. 12-7 Ratio of rms total current to rms symmetrical current for testing Class E controllers.

one phase. During the calibration test, one line-to-line voltage is read while the test power source is open-circuited. Current flowing after closing of the circuit is measured by oscillograph elements in all three phases. Total rms current is determined from the oscillogram traces in a conventional manner by a method described in American Standard C37.5-1953. In adjusting the test circuit, excitation of the test alternator and total impedance are set so as to obtain the 1.6 ratio for the initial half-cycle. The L/R ratio for the external impedance is then so chosen that the d-c component is reduced substantially to zero after four cycles, in order not to impose too severe a duty on the controller during the interrupting test. After the test circuit has been calibrated, the short circuit is removed from the line terminals of the controller and placed across its load terminals. The test cycle consists of three CO operations in which the opening follows immediately after the closing without any purposely delayed action. For Class E1 controllers, the CO operations of a test cycle follow each other at two-minute intervals. For Class E2 controllers, the interval between CO operations is the time to renew the fuses. During a test on a Class E2 controller, welding of the contactor contacts is permissible, therefore they may be inspected and replaced, if necessary, between CO operations. Random switching is used for initiating the test shots, it being representative of the interrupting duty to which a starter may be subjected at an installation.

During an interrupting test, current through the controller and line-to-neutral voltage across the controller are recorded by oscillographs for each phase. The resulting oscillograms serve to determine whether successful interruption has taken place. Also, line-to-line voltage across two phases is measured to prove that recovery voltage is not less than rated voltage of the starter.

Criteria for the successful performance of an interrupting test are as follows:

1. There should be no mechanical damage to starter parts.
2. The starter must be capable, with contactor contacts open, of withstanding rated voltage without dielectric breakdown.
3. The starter must be capable of carrying its rated current and rated voltage for a limited time, although permissible temperature rise of parts may be exceeded.
4. Flame must not be thrown from the enclosure during the test.

The starter need not be able to perform another interrupting test without some overhauling and minor repairs, such as replacement of

CLASS E1 CONTROLLERS 321

contacts. During an interrupting test, slight "spits" of flame may go over to the frame or the enclosure of the starter. Experience has shown that such ground currents of very short duration are not harmful to successful interrupting performance, as long as these spits do not initiate arcs which permit the flow of a substantial amount of ground current. During a test, the starter frame is grounded through a 3-ampere, 5000-volt fuse. If sufficient current flows to ground during a test so as to blow this fuse, the test is considered not to have been successful.

High-voltage controllers may be installed comparatively close to the entrance of high-voltage lines into the plant, and they may thus be subjected to impulse stresses due to traveling waves. To prevent controller breakdown due to such traveling waves, NEMA standards require that Class E controllers be so designed as to withstand an impulse voltage test, in addition to the standard dielectric test of 2000 volts plus $2\frac{1}{4}$ times rated voltage. Impulse tests are made with a 1.5x40 full wave in accordance with page 23 of the *American Standards for Measurement of Test Voltage in Dielectric Tests*, C68.1-1953. The impulse test voltage is 45 kv for controllers rated 2300 volts and 60 kv for controllers rated 4600 volts. The impulse voltage test is independent of the interrupting test, and the controller is not required to meet the impulse voltage test after having been subjected to an interrupting test. The impulse voltage test is to prove the voltage rating of a given controller design; it is not considered a production test.

CLASS E1 CONTROLLERS

A typical Class E1 controller is shown in Fig. 13-7. It is a full-voltage starter enclosed in a sheet-steel cabinet. The lower compartment contains the high-voltage contactor, the disconnect switch, and the fused control transformer. In the front part of the upper compartment, the low-voltage control devices are mounted. A protective sheet, separating the low-voltage devices from the high-voltage bus, has been removed to show the high-voltage bus in the rear of the compartment. Interlocking is provided so that the contactor has to be open before the disconnect switch can be closed, and both the disconnect switch and the contactor must be open before the door can be opened. Conversely, the door must be closed before the disconnect switch can be closed and the contactor energized. This starter has a rated interrupting ability of 100,000 kva at 2300 volts, and 150,000 kva at 4160–4600 volts.

322　MOTOR BRANCH-CIRCUIT PROTECTION

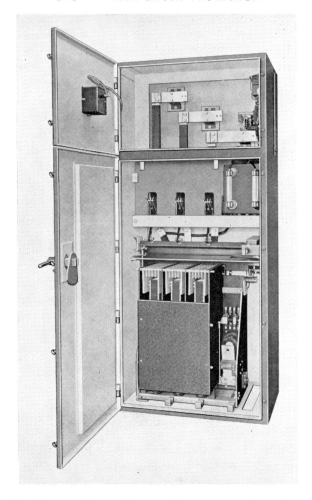

Fig. 13-7　Class E1 controller rated 100,000 kva at 2300 volts and 150,000 kva at 4160–4600 volts interrupting capacity. (Courtesy Square D Company.)

Class E1 controllers can be used on a power system with unlimited short-circuit capacity by installing current-limiting reactors in the main power bus of the starter. These reactors are so selected that they reduce the available short-circuit capacity to a value commensurate with the interrupting ability of the controller. Class E1 controllers with current-limiting reactors are available under the name Valimitor *

* Registered trade mark of the Square D Company.

CLASS E2 CONTROLLERS

starters. Figure 14-7 shows the lower compartment of a Valimitor starter, containing the line contactor and disconnect switch. On the left side, the air-core reactors can be seen mounted on the side sheet of the enclosure.

CLASS E2 CONTROLLERS

Most Class E2 controllers built today use current-limiting high-voltage fuses as a short-circuit interrupting means. Such fuses are built especially for motor branch-circuit protection. They contain silver ribbons as fusible links which are imbedded in quartz sand and enclosed in a composition tube of high mechanical strength. The mass

Fig. 14-7 Class El Valimitor controller with current-limiting air-core reactors. (Courtesy Square D Company.)

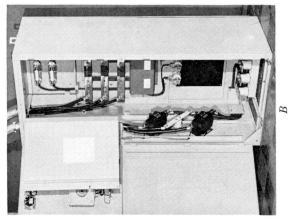

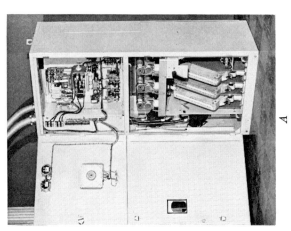

Fig. 15-7 Class E2 Limitamp controller rated 150,000 kva at 2300 volt and 250,000 kva at 4160–4600 volts interrupting capacity. *A*. Door open with contactor unit in place. *B*. Contactor assembly withdrawn. *C*. Close-up of contactor assembly with current-limiting fuses.

of the fusible links is so selected that the fuses have sufficient lag so that they will not melt on inrush current when starting squirrel-cage or synchronous motors. On faults beyond the interrupting ability of the line contactor, they open sufficiently fast to keep the let-through energy low.

A Class E2 Limitamp * controller is illustrated in Fig. 15-7. A is a view of the controller with the doors open. The upper compartment contains low-voltage control devices which are mounted on a swinging steel panel behind which the high-voltage bus and incoming power connections are located. The lower compartment contains the high-voltage assembly of line contactor, disconnect device, and high-voltage fuses. This whole assembly is plugged into stationary contacts in the enclosure, and it can be withdrawn and rolled out of the compartment for easy maintenance. The door of the high-voltage compartment is interlocked so that it must be closed before the contactor can be energized. Conversely, the contactor must be open before the disconnect can be opened, and the disconnect must be open before the door can be opened. The upper door to the low-voltage compartment can be opened at any time, so that relays may be adjusted with the motor running.

B shows the controller with contactor assembly withdrawn from the compartment and disengaged from the stationary contacts. In the upper compartment, the low-voltage panel has been swung out, showing the location of the high-voltage bus. C is a close-up of the high-voltage assembly, showing the details of the disconnect device and current-limiting fuses assembled on their supports on top of the contactor structure. The controller shown has an interrupting rating of 150,000 kva at 2300 volts, and 250,000 kva at 4160 to 4600 volts.

BIBLIOGRAPHY

1. National Fire Protection Association, *National Electrical Code*, 1959 edition.
2. B. W. Jones, "The Behavior of Industrial Control Devices under Short Circuit," *General Electric Review*, October 1933.
3. D. C. Prince and E. A. Williams, "The Current-Limiting Power Fuse," *Transactions AIEE*, January 1939.
4. B. W. Jones, "The Degree of Short-Circuit Protection Afforded Small Low-Voltage A-C and D-C Starters by Means of Fuses and Circuit Breakers," *Transactions AIEE*, August 1944.
5. H. L. Rawlins and J. M. Wallace, "Modern High-Voltage Fuses," *Westinghouse Engineer*, September 1944.

* Registered trade mark of the General Electric Company.

6. C. L. Schuck and E. W. Boehne, "Performance Criteria for Current Limiting Power Fuses, Parts I and II," *AIEE Paper* 46-170.
7. W. J. Herziger, "Fused Starters Protect High-Voltage Equipment," *Allis-Chalmers Review*, 2nd and 3rd Quarter 1946.
8. G. W. Heumann, "High-Voltage High-Interrupting Capacity A-C Motor Controllers," *General Electric Review*, April 1950.
9. C. L. Schuck, "New Low-Voltage Power Fuse of High Interrupting Capacity," *General Electric Review*, July 1950.
10. R. P. Ballou, "Fuses versus Circuit Breakers," *Electrical Manufacturing*, February 1951.
11. T. B. Montgomery and T. H. Bloodworth, "High-Voltage Motor Controllers Coordinated with Distribution Switchgear," *Proceedings AIEE 1951*, Section T1-252.
12. K. S. Kuka, "Fuses and Air Circuit Breakers for Short-Circuit Protection," *Iron and Steel Engineer*, May 1951.
13. B. J. Dalton, "New Control for High-Voltage Industrial Motors," *General Electric Review*, May 1951.
14. G. W. Heumann, "High-Voltage A-C Motor Controllers," *Electrical Engineering*, January 1952.
15. J. W. Gibson, "The Application and Standardization of High Rupturing Capacity Current-Limiting Fuses," *AIEE Paper* 53-129.
16. J. C. Lebens, "Coordinated Fuse Protection for Low-Voltage Distribution Systems in Industrial Plants," *AIEE Paper* 54-9.
17. A. H. Powell and C. L. Schuck, "Ribbon Elements for High Voltage Current Limiting Fuses," *AIEE Paper* 55-217.
18. G. W. Heumann, "How to Select Today's High-Voltage A-C Motor Controllers," *Power*, October 1955.
19. C. A. Lister, "Interruption Tests on High-Voltage Air-Break Contactor," *AIEE Paper* 56-31.
20. W. C. Huening, "Low Voltage Switching and Protective Device Characteristics," *AIEE Paper* 56-129.
21. G. W. Heumann, "Applying Industrial Controllers for Safe Operation," *AIEE Paper* 56-269.
22. E. W. Davis, "Application of Low Voltage Switchgear for Motor Control," *AIEE Paper* 56-327.
23. G. W. Heumann, "How to Engineer Motor Branch Circuits," *Mill and Factory*, Part 1, April 1956, Part 2, June 1956.
24. K. S. Kuka, "Short Circuit Protection of Low Voltage Motors and Starters," *Iron and Steel Engineer*, July 1956.
25. P. C. Jacobs, "Current Limiting Fuses—Their Characteristics and Applications," *AIEE Paper* 56-772.
26. M. S. Carlson and W. H. Edmunds, "Coordination of Current Limiting Fuses and Low Voltage Air Circuit Breakers," *AIEE Paper* 56-773.
27. W. H. Edmunds, "A New High Interrupting Capacity Low Voltage Circuit Breaker," *AIEE Paper* 56-774.
28. C. A. Lister, "High Interrupting Capacity Controllers for 2300–4600 Volt Motors," *Iron and Steel Engineer*, February 1957.
29. S. R. Durand and T. Bellinger, "Operating Characteristics of High Capacity Air-Break Contactors on Industrial Power Systems," *AIEE Paper* 57-266.

PROBLEMS

30. E. M. Fitzgerald and V. N. Stewart, "High Capacity Current Limiting Fuses Today," *AIEE Paper* 58-1188.
31. H. J. Wilt, "Circuit Factors in Motor Protection," *Electrical Manufacturing*, June 1959.

PROBLEMS

1. In an industrial distribution system having a no-load voltage of 460 volts, the impedance of the step-down transformer is 0.015 ohm and the impedance from the transformer to the point of controller installation is 0.004 ohm. Determine the following:
   (a) The available symmetrical short-circuit current.
   (b) The required interrupting rating of a circuit breaker to protect the above circuit.

2. A line of air-circuit breakers with thermal-instantaneous overload trips has trip ratings of 15, 20, 30, 40, 50, 70, 100, 125, 150, 175, 200, 225 amp. Select trip ratings for circuit breakers for use with the following motors:
   (a) Single-phase, 1 hp, 115 volts, 13 amp.
   (b) Three-phase, squirrel-cage, 5 hp, 440 volts, 7.5 amp, Code Letter G, with full-voltage starting.
   (c) Three-phase, synchronous, 60 hp, 440 volts, 80 amp, Code Letter H, with autotransformer starting.
   (d) Three-phase, wound-rotor, 50 hp, 220 volts, 138 amp.
   (e) Three-phase, squirrel-cage, 30 hp, 440 volts, 40 amp, driving centrifuge, with extra-heavy starting condition.

3. Select fuses for branch-circuit protection of the following motors:
   (a) Direct-current, 2 hp, 230 volts, 8.5 amp.
   (b) Single-phase, 3 hp, 230 volts, 17 amp.
   (c) Three-phase, squirrel-cage, 5 hp, 440 volts, 8.5 amp, Code Letter F, with full-voltage starting.
   (d) Three-phase, squirrel-cage, 5 hp, 440 volts, 8.5 amp, Code Letter G, with autotransformer starting.
   (e) Three-phase, squirrel-cage, 3 hp, 220 volt, 9 amp, driving centrifuge, with extra-heavy starting condition.

4. A 4160-volt distribution system is connected to a 7500-kva transformer having an impedance of 0.035 per unit. The impedance of the feeder to the point of controller installation is 0.01 per unit:
   (a) What is available symmetrical short-circuit kva at point of controller installation?
   (b) What class of controller would you recommend, and what is its rated interrupting capacity?

5. A 2300-volt distribution system is connected to a 2000-kva transformer having an impedance of 0.04 per unit. The impedance of the feeder to the point of controller installation is 0.01 per unit:
   (a) What is available symmetrical short-circuit kva at point of controller installation?
   (b) What class of controller would you recommend, and what is its rated interrupting capacity?

INDEX

accelerating time, 5
   torque, 5
      average, 6
acceleration, 5, 89, 179
   current-limit, 98
   definite-time, 102
   magnetic control of, 97
   secondary-frequency, 100
across-the-line starter, 20
adjustable-speed motor, 86
adjustable-voltage d-c drive, 158, 167
air circuit breaker, 26, 302
Allen-Bradley Co., 193, 282, 284, 287
Allis-Chalmers Mfg. Co., 193, 210
ambient-compensated relay, 261
ambient temperature, 244, 261, 265, 303
American Standard C37.5, 320
   C68.1, 321
   Y32.2, 223
amortisseur winding, 179
amplifier, 222
analog device, 240
AND function, 218
antikiss circuit, 35
armature reaction, 11
asymmetrical current, 293
automatic reset (of overload relay), 253
autotransformer starter, 47, 50
auxiliary function, 222, 227
available short-circuit current, 296

bias, 240
bistable device, 238, 240
braking-lowering quadrant, 3
braking torque, 6
breaker-fuse combination, 310
bridge, crane, 116
buffer resistor, 69

cam switch, 96
circuit breaker, 295, 302
   air, 26, 302
   current-limiting, 310
   fuse combination, 310
   instantaneous-trip, 304, 307
   thermal-trip, 303
   time-limit, 302
circuit (logic) design, 225
Clark Controller Co., 193
Class A controller, 293, 317
Class C controller, 308
Class E controller, 317
Class E1 controller, 317, 321
Class E2 controller, 317, 323
closed-circuit transition, 18, 52, 65
code letter, 14
combination light and power system, 45, 58, 274
combination starter, 25, 308
compelling relay, 74
consequent-pole winding, 69
constant-horsepower motor, 69
constant-speed motor, 11, 177
constant-torque application, 4
constant-torque motor, 69
contact reliability, 237
control center, 313
control-circuit device, 216
control-circuit fuse, 24
control sequence chart, 226
control transformer, 22, 33
conversion factor (for overload relays). 265
coordination (of protective devices). 304
Cordon circuit breaker, 310, 312
crane classification, 114
   control, 115

329

330　INDEX

crane classification, service, 114
current-limit acceleration, 98
current-limiting action, 299
current-limiting circuit breaker, 310
current-limiting fuse, 299
current-limiting reactor, 315, 322
Cutler-Hammer, Inc., 135, 193
cycling, 31

damper winding, 179
d-c component, 16, 294
d-c dynamic braking, 43
  lowering, 138
decision-making element, 216
decrement, 16
deep-well pump, 28
definite-time acceleration, 102
DELAY function, 218, 230
delta-wye transformer, 279
demagnetizing action, 189
digital device, 240
discharge resistor, 181
distributed starting winding (of synchronous motor), 182
distribution transformer, 278
  delta-delta-connected, 278
  delta-wye-connected, 279
  wye-delta-connected, 279
  wye-wye-connected, 278
double cage, 12
double-range system, 162
double-wye resistor, 107
doubly-fed induction motor, 165
drum switch, 95
dual-voltage motor, 18, 251
duplex controller, 122
dynamic braking, 109, 182
  hoist controller, 125
  resistance, 182
  torque (of synchronous motor), 182

economic comparison of starters, 67
Edison Electric Institute (EEI), light flickers, 45
Electric Machinery Mfg. Co., 193
electronic stepless control, 146
elevator, 285
eutectic solder, 261
exciter, 177

face-plate controller, 93
fan load, 4, 159
fast-trip relay, 259, 262, 269
favorable angle, 186
field application, by polarized field frequency relay, 192
  by slip-frequency relay, 196
  by timing relay, 191
field coil, 177
field control panel, 198
field removal, 202
field rheostat, 190
flux-decay relay, 106, 196
flux rotation, 9
foreign-built motors, 277
frequency-changer system, 164
friction load, 3
full-voltage starter, 20, 181
  standard connections of, 20
fuse, control-circuit, 24
  current-limiting, 299
  high-voltage, 323
  low-voltage, 299

gate, 223
general-purpose motor, 11, 245
  starter, 15
generated rotor voltage, 9
graduated plugging controller, 126

hand reset (of overload relay), 253
Harnischfeger Corp., 146
heater table, 264
hermetic compressor, 251, 269
  motor, 247, 269
high-inertia load, 247
high-slip motor, 14
high-starting torque motor, 12
high-voltage controller, 298
  motor, 33
  starter, 33
hoist, balanced, 124
  crane, 116

Ideal Electric Co., 193
impact load, 206, 270, 272
impulse voltage test, 321
inching, 29
incomplete-sequence relay, 207

INDEX 331

increment starter, 55
indexing, 31
index point, 31
induction controller, 147
induction motor, 8
inertia, moment of, 5
inherent protection, 247
inherent protector, 248
instantaneous overcurrent relay, 272
instantaneous overload relay, 272
  setting (of protective device), 17
  trip circuit breaker, 304, 307
insulation life, 245
interlock, electrical, 37
  mechanical, 36
  potential, 38
interrupting capacity, 298
ITE Circuit Breaker Co., 310

jogging, 29, 35
jog pushbutton, 30
  relay, 30
Joint Industry Conference (JIC), control transformer, 22

Klixon protector, 249
Korndorfer connection, 52
Kraemer system, 159

laminated poles, 177
light flicker, 45
Limitamp controller, 324
limit switch, 120
linear motion, 5
line contactor, 20
load angle, 187
  brake, 116, 119, 124
  characteristic, 3
  fan, 4
  friction, 3
Load-O-Matic hoist control, 153
locked-rotor current, 14, 183
lock-in-step control (of selsyns), 169
logic, 216
  element symbols, 217
  function, 215
loop (of a logic element), 238

machine function, 223
machine-tool duty, 4

magnetic amplifier, 238
  self-saturating, 239
magnetic control of acceleration, 97
magnetic relay, 237
magnetizing current (of induction motor), 9, 40
manual controller, 93
master switch, handle throw, 118
  spring-return, 20
maximum torque, 12, 40, 179
mechanical interlock, 36
  life, 237
  timer, 104
MEMORY function, 220, 231
Metals and Controls Corp., 249
minimum slip, 182
molded-case circuit breaker, 26, 305, 312
moment of inertia, 5
motor branch circuit, 292
  protection, 292
motor burnout, 244
  circuit switch, 26, 299
  control center, 313
  heating curve, 245
  protection, 244
multiple-wye resistor, 108
multispeed motor, 69
  starter, 71

National Electrical Code (NEC), circuit breaker, 302
  code letter, 14, 305
  disconnect means, 24, 299
  "four times" rule, 304
  fuse and circuit-breaker ratings, 304
  fuse dimensions, 299
  motor branch-circuit protection, 292
  number of circuit-breaker trips, 304
  number of fuses, 304
  number of overload relays, 254
  overload protection, 244
  overload relay heaters, 264
  phase-reversal protection, 285
  short-circuit protection, 24, 309
  three overload relays, 281
National Electrical Manufacturers' Association (NEMA), Class E controller standards, 318
  code letter, 14

INDEX

National Electrical Manufacturers' Association (NEMA), controller classification, 317
crane classification, 114
crane resistors, 121
crane standards, 115, 116
fuse standards, 299
impulse voltage test, 321
logic element symbols, 217
master switch handle throw, 118
motor designs, A, B, C, D, F, 12
number of accelerating contactors, 111
part-winding starter standard, 62
starter connections, 20, 36
National Machine Tool Builders' Association (NMTBA), control transformer, 22
network starter, 56
normal starting-torque motor, 12
NOT function, 218, 229
number of accelerating contactors, 111
number of overload relays, 254
number of phases, 8
number of poles, 8, 177

offset, 16, 294
ohmic drop exciter, 162
open-circuit transition, 18, 52, 65, 70
OR function, 218, 229
overcurrent relay, 272
overheating, 244
overload, 244
  protection, 244
  relay, 252
    ambient-compensated, 261
    bimetal, 255
    correlation of, 305
    fast-trip, 259, 262, 269
    induction, 259
    instantaneous, 272
    medium-trip, 260, 262
    number of, 254
    resistance-type, 256
    self-protecting, 306
    slow-trip, 260, 262
    solder-film, 261
    thermal, 252
overshoot (of overload relay), 267

part-winding starting, 56, 181, 304
phase-failure protection, 281
phase-reversal protection, 285
phases, number of, 8
phase sequence, 9
pilot-motor operating mechanism, 97
planer table, 231
plugging, control, 41, 109, 119
  of induction motor, 39, 109
  switch, 41
  of synchronous motor, 182
  torque, 40, 110, 182
plug-stop control, 41
pneumatic conveyor, 4
  timing interlock, 105
points (of a starter), 90
polarized field frequency relay, 192, 202
pole changing, 86
  contactor, 74
polechanging motor, 11
poles, number of, 8, 177
polyphase motor, 9
potential interlocking, 38
power factor, correction, 189
  field removal relay, 204
  of synchronous motor, 176, 189, 203
power output element, 216
power selsyn, 167
primary-reactor starter, 48, 56, 147
primary-resistor starter, 48, 54
primary winding, 8
progressive acceleration, 74
protective features, 244
pull-in torque, 187
pull-out, characteristic, 188
  protection, 205
  relay, 202
  torque, 179, 187

quadrants of motor performance, 2

reactor control, 142
receiver (selsyn), 167
reconnection (of motor to line), 18
recovery voltage, 295
Rectiflow drive, 160
reduced-inrush starting, 44
reduced-voltage starter, 47, 181
regenerative braking, 39, 109

INDEX

reject-type fuse clips, 300
reluctance (of rotor path), 183
reluctance motor, 186, 210
replica-type relay, 252
reset, automatic, 253
  hand, 253
resistance temperature detector (RTD), 248
resistance-type heater, 256
resistor application data, 91
  classification, 91
  layout, 92
retardation, 5, 179
retarding torque, 7
reversing, 35, 108
  controller, 109
  plugging controller, 119, 144
rheostat ohms, 83
rotating flux, 9
rotor, 8, 176
  frequency, 12
  impedance, 12
  voltage, 10

saliency, 183
salient poles, 176
saturable reactor, 142, 147, 238
saturable transformer, 147
Scherbius system, 162
secondary current, 11
  frequency acceleration, 100
  power conversion, 158
  reactor control, 142
  resistance, 10, 83
  voltage (of induction motor), 9
self-protecting overload relay, 306
selsyn, 167
semimagnetic controller, 93
series lockout, 98
service factor, 257, 264
  life, 245
short circuit, 292
  current, 294
    available, 296
  impedance, 295
    power factor, 295
  protection, 293
signal converter, 223
single-phase dynamic lowering, 132

single-phasing, 75, 252, 272
single-range system, 162
slip, 9, 82
  frequency relay, 196
  minimum, 182
  power, 158
  ring, 82, 176
Slipsyn controller, 200
solenoid brake, 120
speed-load curve, 116
speed-torque curve, 1, 116
Spencer disk, 248
Square D Co., 193, 322, 323
squirrel-cage motor, 8
  protective relay, 196, 206, 272
  winding, 8, 179
  protection, 207
stall time, of induction motor, 257
  of synchronous motor, 181
standstill voltage (of wound-rotor motor), 83
starting, of squirrel-cage motors, 20
  of synchronous motors, 180
  of wound-rotor motors, 89
starting current, 14
  resistor, 91
  torque, 12, 91, 180
static switching, 215, 237
stator, 8, 176
steps (of a starter), 90
superimposed winding, 69
switching transient, 15
symmetrical components, 75, 129
  current, 293
synchronizing, of selsyn, 169
  of synchronous motor, 179, 182
synchronizing torque, of selsyn, 167
  of synchronous motor, 183
synchronous motor, 176
  controller, 190
  speed, 9, 176
synchrotic, 167
Synduction motor, 210

tapped field resistor, 190
thermal relay, 208
  overload relay, 252
trip (of circuit breaker), 302
three-point control circuit, 20

three-step method (of selsyn starting), 170
Thrustor brake, 120
thyratron, 146
time constant (of motor winding), 16
tolerance (of overload relays), 256
transient, switching, 15
transient current peak, 17
transmitter (selsyn), 167
transistor, 238, 241
 amplifier, 241
transition, closed-circuit, 18, 52, 65
 open-circuit, 18, 52, 65, 70
Tri-Pac circuit breaker, 310
trolley, crane, 116
turn ratio (of induction motor), 9
two-phase motor, 24
two-point control circuit, 23
two-step method (of selsyn starting), 170

unbalanced rotor connection, 88
unbalanced stator voltage lowering, 135
unbalance of primary voltage, 129
undervoltage protection, 26, 121
 time delay, 27, 34
undervoltage release, 26
Underwriters' Laboratories, Inc. (U/L),
 fuse test, 299
 motor-protector tests, 249, 251
 overload relay heaters, 264
 starter tests, 15
unfavorable angle, 186

Valimitor controller, 322
variable-torque motor, 69
V-curves, 189

Westinghouse Electric Corp., 153, 160, 200, 310
wind tunnel, 159
wound-rotor motor, 8, 81
 protection, 270
 starter, 111
wye-delta starting, 63
wye-delta transformer, 179